Symbols

a	Specific surface area of supporting media (cm^{-1})
J	Flux of rate-limiting substrate into biofilm (mg/sqcm/day)
J_{avg}	Average substrate flux in biofilm (mg/sqcm/day)
k	Maximum specific rate of substrate use (day^{-1})
S_0	Concentration of rate-limiting substrate in influent (mg/cc)
S_w	Effluent substrate concentration (mg/cc)
Θ	Empty-bed hydraulic detention time (hr)
Θ_c	Mean Cell Residence time or Solid Retention Time (day^{-1})
μ	Specific growth rate of biomass (day^{-1})
b_s	Biomass loss rate due to shearing from biofilm (day^{-1})
b_d	Biomass decay coefficient (day^{-1})
b_t	Total biomass loss rate from biofilm (day^{-1})
K_s	Half-velocity coefficient (mg/cc)
Y	Biomass yield coefficient (mg/mg)
X_{hybrid}	Concentration of bomass in hybrid bioreactor(mg/cc)
p	Porosity of hybrid reactor
V	Empty-bed volume of hybrid reactor (m^3)
X	Concentration of suspended biomass in hybrid reactor(mg/cm^3)
S_f	Substrate concentration at any point in the biofilm (mg/cm^3)
X_f	Active biomass density within the biofilm(mg/cm^3)
D_f	Molecular diffusion coefficient of the substrate in the biofilm(cm^2/day)
S_{min}	Minimum concentration of rate-limiting substrate at biofilm-attachment surface(mg/cm^3)
L_e	Effective biofilm thickness
L_f	Total biofilm thickness
Q_w	Waste sludge flow rate (m^3/hr)
X_r	Biomass concentration in waste sludge (mg/cc)

Chapter 1
Introduction

1.1 Background

Due to the rapid increase in population growth, urbanization, industrialization and more luxurious lifestyles, huge amount of pollutants are produced day by day and discarded into the water system. Billions of gallons of wastewater are produced every day from domestic and industrial sources. Due to the adverse effect of different pollutants of wastewater on the surrounding water environment, wastewater treatment becomes inevitable nowadays. The treatment of wastewater should be adequate enough to satisfy the discharge standard so that potential of the final effluent remains within natural purification capacity of the watercourse. Biological treatment method plays an important role in removing biodegradable carbonaceous matter from domestic and industrial wastewater. Out of various biological methods activated sludge process (ASP) and its several modifications are widely being used throughout the world. Apart from that, attached-growth fixed-film system can also be used for the biological treatment of both domestic and industrial wastewater. Among these two, activated sludge process exhibited better flexibility in operation to improve the effluent quality. The ASP system can also be used for simultaneous carbon oxidation and nitrification in its modified configurations.

However, activated sludge process has also experienced some inherent problems while functioning in the wastewater treatment plants. The major problems encountered by this process are to maintain the uniform biomass concentration in the aeration tank, increased volume of tank, incapability of the biomass to resist the shock effect due to sudden increase in the organic load in wastewater, poor settleability in secondary clarifier, inadequacy to withstand toxic and inhibitory substances, high power inputs toward pumping for recirculation, etc. Even, for the treatment of low strength wastewater like municipal wastewater, biomass in the activated sludge process is susceptible to washout if there is no recirculation of biomass To alleviate these limitations and operational problems of ASP, the current practice is to integrate the fixed-film media into activated sludge reactor, which is emerged as integrated fixed-film activated sludge system, also known as aerobic hybrid bioreactor.

S. Sarkar and D. Mazumder, *Fixed Bed Hybrid Bioreactor*, Green Energy and Technology, https://doi.org/10.1007/978-981-33-4546-1_1

Although the integrated fixed-film activated sludge system was first developed in the USA in 1980, a long time back, there is a very few research work done so far in the development of mathematical model considering the concurrent growth of both suspended biomass and attached biomass, which is required for the process design of the hybrid bioreactor. This is because the biological reactions involving the conversion of the substrates by both suspended- and attached-growth biomass in a concurrent way are highly complex and especially the solution of biofilm model is much more tedious and cumbersome compared to that of suspended-growth process. Even where the competition for rate-limiting substrates between two growths (both suspended and attached) was simultaneously considered, no unique accurate and simplified solution was derived from the steady-state substrate mass balance and biomass balance equations.

A simplified mathematical model (with computer programming and excel worksheet) for fixed-bed hybrid bioreactor thus finds its relevance for predicting the reliable outputs for the sake of process design of the reactor. Accordingly, a simplified model of an aerobic hybrid bioreactor is developed considering simultaneous growth of both the suspended and attached biomass in a competitive manner. The model is based on steady-state carbonaceous substrate as well as biomass mass balance for both suspended and attached growths along with substrate mass transport into the biofilm. The analytical solution involved determination of the average flux along with exiting substrate concentration and determination of effective biofilm thickness by Runge–Kutta method. A computer program is developed in Fortran language and the model has been validated with both the existing methods of standard literatures and the experimental results. Monod kinetic relationship is used in the present model considering no inhibition in remaining carbonaceous organic matter from the municipal wastewater.

1.2 Historical Findings

Integrated film activated sludge process was first developed in the USA in 1980, where biofilm was inserted into the aeration tank [1, 2]. There are two types of attached media entrapped in the aeration tank like "dispersed media," i.e., movable type and "fixed media," i.e., sheet media or fabric media fixed in place in the aeration tank. IFAS technology provides for additional biomass within a wastewater treatment facility in order to meet more stringent effluent discharge standard or increased organic loadings without the direct need for additional tank [3]. The hybrid bioreactor process was applied as a retrofitting of the existing ASP system without expanding the volume of the reactor. Hybrid bioreactor technology has been incorporated into municipal and industrial wastewater facilities for upgrading or retrofitting plants in many variations of suspended-growth systems. In case of upgraded plants, additional treatment capacity can be obtained without increasing the volume of the existing tank. The secondary treatment processes were upgraded to integrated fixed-film activated sludge process by adding the media into its existing basin in Broomfield wastewater treatment plant for improving the biological nutrient removal process [4].

After the IFAS process, the LINPOR process was developed as a modified activated sludge system, where highly porous suspended plastic foams cubes were used as a movable media in the aeration tank. The LINPOR process produced a treated effluent quality far better than the conventional ASP process [5, 6].

The suspended carrier biofilm process (Kaldenes moving bed biofilm reactor) was applied in the existing treatment facilities [7–10, where high-density polyethylene dispersed biofilm media was used as biofilm attachment surface. This process resulted in efficient carbon removal and improved nitrification and denitrification in the system. For a process design of the hybrid bioreactor, a user-friendly mathematical modeling is required for finding out the relevant outputs. The first mathematical model of the hybrid bioreactor was developed by Lee [11]. Thereafter, one computer program was developed for hybrid bioreactor for removing soluble COD and nutrients [12]. Plastic nets as an attached surface were inserted into the aeration tank and the mathematical model of the reactor was accordingly developed [13, 14]. One mathematical model was developed for a steady-state biofilm activated sludge reactor to calculate the substrate flux in the biofilm under substrate limiting condition [15]. Effort has also been made to develop a simplified mathematical model for designing the steady state biofilm activated sludge reactor under limiting substrate condition [15]. Earlier one activated sludge model (ASM2d) was developed for biological phosphorus removal with simultaneous nitrification–denitrification in the activated sludge process [16]. The said model was further extended to a steady-state IFAS model by Boltz et al. [17] using the input taken from biofilm modeling techniques [18]. Eventually, a simplified mathematical model was proposed to provide an accurate tool for describing the steady-state suspended-growth biofilm system in the treatment of municipal wastewater [19].

The present model developed by the author is not only found very simple, fast and accurate method in determining the output parameter, but also it can calculate the effective biofilm thickness unlike other solution models. Actually, it is the effective biofilm thickness which contains the biomass actively metabolizing the substrate beyond which the substrate flux ceases to get utilized further. Moreover, from the effective biofilm thickness it can be ascertained whether the biofilm is a shallow or deep biofilm. Apart from that, the solution model employed the kinetics of suspended- and attached-growth biomass in integrated manner considering their simultaneous growth. Unlike other existing models, average substrate flux (Javg) is considered in the present mathematical model, considering a variation in substrate flux from biofilm layer to layer on account of substrate gradient. The results of performance study on the laboratory-scale hybrid bioreactor also established the accuracy of the present solution model. Such experimental validation can be carried out with different types of wastewater of varying strength for examining its versatility. All these may lead to exploring a generalized mechanism for process design of a hybrid bioreactor.

In case of municipal wastewater like low strength wastewater, suspended biomass is subjected to washout if there is no recirculation of biomass. Thus, the addition of biofilm attached media into the aeration tank for maintaining the proper biomass in the reactor can reasonably be thought for the treatment of municipal wastewater.

1.3 Brief Overview on Past Experience

It has been found from the past experience that the hybrid bioreactor process has significantly improved the efficiency of biological treatment of slowly biodegradable substance and has been widely applied to municipal wastewater treatment in recent years. In the past, the hybrid bioreactor technology has been incorporated into municipal wastewater treatment as a new plant installation as well as for upgradation/retrofitting of the overloaded plants in many variations. In case of new plant installation, the reactor volume gets reduced due to additional attached biomass. For the plant upgradation, enhanced treatment capacity can be obtained without increasing the volume of the existing tank. Hybrid bioreactor was used for studying the nutrients removal from municipal wastewater [19]. It was found that the hybrid bioreactor ensured a better removal than the purely attached biofilm system. IFAS process was also installed at Maryland municipal wastewater reclamation facility, where rope like ringlace IFAS media of 30,000 m were installed in a volume of 475 cum. Nitrification in the IFAS section was found more than three times that in the control section without media [9]. A sponge type IFAS media was used at Moundsville wastewater treatment plant, USA in 3 years of full-scale operation. This is a combined suspended- and attached-growth process, where polyurethane foam media was incorporated into ASP process increasing the equivalent MLSS concentration level. The attached and suspended MLSS were maintained between 6500–12,000 mg/l and 900–1700 mg/l, respectively.

The efficiency in removal of BOD, nitrification and denitrification was found 95%, 70–90% and 40–60%, respectively, at a HRT 50–90 min [20].

A study was conducted to see the effect of attached biomass on the filamentous growth of suspended biomass in the nitrification process and found that nitrification became independent of SRT of suspended biomass, since nitrifying bacteria are predominantly attached to the support materials. In this case, dispersed plastic foam particles and fully submersed plastic disks were used as a supporting media for attached biomass [21]. In a cold region, treatment of domestic wastewater was done by a hybrid activated sludge biofilm process, which contained both suspended biomass and biofilm, usually referred as integrated fixed-film activated sludge (IFAS) process, created by introducing plastic elements as biofilm carrier media into a conventional activated sludge reactor. Pilot-scale study was carried out at the Department of Hydraulic and Environmental Engineering at the Norwegian University of Science and Technology in Trondheim. The results revealed that this kind of reactor can efficiently be used for the upgrading of conventional activated sludge plant for achieving a good quality of effluent [22]. Sriwiriyarat et al., [23] also made his research study treating a weak municipal wastewater with IFAS using sponge media at different aerobic mean cell residence times for evaluating his model performance with the experimental data and found satisfactory results in removal of carbonaceous organic matter and nitrification.

1.4 Work Executed as Per Simplified Model Developed by Author

A simplified mathematical model has been developed each for fixed biofilm reactor, completely mixed biofilm reactor and aerobic fixed-bed hybrid bioreactor. After developing the model of fixed-bed hybrid bioreactor, an analytical procedure for determining the kinetic coefficients of hybrid bioreactor was developed. The development of the kinetic coefficients was based on a set of straight lines derived from the expression of mathematical model. A series of graphs were formulated based on the straight line equations and the kinetic coefficients were determined either from the intercept of *x*- and *y*-axis or from the slope of those graphs.

Thereafter, analytical validation of the mathematical model with one or more sets of available data in the past existing models cited in literatures was done with a same set of given kinetic coefficients in each case as given in the literature. After the analytical validation of the mathematical model, a laboratory-scale fixed-bed hybrid bioreactor was fabricated both for conducting the semi-batch and continuous study with both synthetic carbonaceous and real municipal wastewater. The reactor was set up for performance study of the fixed-bed hybrid bioreactor, for determination of kinetic coefficients from semi-batch study and for validation of the experimental data with the model outputs under continuous study using both municipal and synthetic wastewater.

Once the reactor set up was complete, initiation was taken for biomass acclimation in the suspended-growth system, i.e., in a 15-L PVC Jar associated with two nos of aqua pumps for aeration. Periodic monitoring for biological acclimation was performed in two phases, Phase A and Phase B. Phase A was performed from the summer period and Phase B was done in the monsoon period. The batch study in two phases was conducted to see the effect of seasonal fluctuations on biological acclimation. Tracer response study is the next step to assess the hydraulic performance of the full-scale reactor used for the treatment of municipal wastewater. The assessment of the flow patterns showed that the reactor was completely mixed in characteristic. The use of tracers (in this case, lithium chloride) for developing the residence time distribution curves is one of the simplest and successful method for assessing the hydraulic performance of the full-scale reactors.

After identifying the hydraulic characteristics of the full-scale reactor, preliminary study can be conducted with synthetic carbonaceous wastewater to check the performance comparison between fixed-bed hybrid bioreactor and plain suspended-growth reactor, i.e., ASP system. If the two reactors can be run in the same operating conditions of input COD and MLSS at same batch period, it will be seen that the performance of the hybrid bioreactor is always better than the suspended one.

The detailed experimental study can be conducted both under semi-batch and continuous mode with synthetic carbonaceous wastewater as well as raw municipal wastewater. The significance of this experimental studies is to observe the removal performance of the reactors and hence to validate the experimental data with the model outputs. It is normally seen that the model output results are within ±10% variation of the experimental outputs. Generally after the experimental validation of the mathematical model, the process design of hybrid bioreactor should be accomplished to determine its volume, physical dimension and other operational requirements including oxygen requirement for any targeted effluent substrate concentration.

For carrying out the experiments in the laboratory, Protein Concentration of the attached biomass sample can be determined following the Lowrys' Method [24] modified by Herbert et al. [25]. The protein concentration is determined to calculate the biomass density of the attached surface in the hybrid bioreactor. The thickness of biofilm can be measured in scanning electron microscope (SEM) following the standard protocols.

The isolation of biomass from both suspended and attached growth can be performed following the procedures like sterilization, serial dilution, augur plating and spreading, and finally gram staining. Finally, characterization of above biomass can be done using 16Sr DNA protocol after PCR analysis.

References

1. Sen D, Randall CW (1994)/1990 Performance of fixed film media integrated in the activated sludge reactors to enhance nitrogen removal. Water Sci Technol 30:13–24
2. Eckenfelder WW Jr (2000) Industrial water pollution control, 3rd edn. McGraw-Hill, Inc., Singapore
3. Campbell H, Schnell A (2003) Upgrading activated sludge systems using free floating plastic media[online]. Available from www.hydroxyl.com/papers/waterdown/UpgradeAS.pdf
4. Rutt K, Seda J, Johnson CH (2006) Two years case study of integrated fixed film activated sludge at Broomfield Co WWTP. Proceedings of the Water Environment Federation, sesson 10, pp 225–239
5. Morper MR (1994) Upgrading of activated sludge system for nitrogen removal by application of the Linpor-CN process. Water Sci Technol 29:167–176
6. Gilligan TP, Morper M (1999) A unique process for upgrading conventional activated sludge systems for nitrogen removal (online). Paper presented at NE WEA, October 1999, available from
7. Harvey PJ, Siviter CL (1999) Use of the suspended carrier process to upgrade wastewater treatment facilities [online]. Available from https://www.purac.net/downloads/articles/1999%20NZWWA%20Conference%20Paper%Kaldnes_pdf. Accessed on 10 Aug 2003
8. Ødegaard H, Rusten B, Westrum T 1994 A new moving bed biofilm reactor applications and results. Water Sci Technol 29:157–165
9. Rantanen P, Valve M (2003) A hybrid process for biological phosphorus and nitrogen removal-pilot plant experiments [online]. Available from www.vyh.fi/eng/syke/ppd/ws/biorev.htm
10. Dalentoft E, Thulin P 1997 The use of the Kaldnes suspended carrier process in treatment of wastewaters from the forest industry. Water Sci Technol 35:123–130 https://doi.org/10.1016/S0273-1223(96)00923-7
11. Lee C-Y (1992) Model for biological reactors having suspended and attached growths. J. Environ Eng 118(6):982–987

12. Sriwiriyarat T (1999) Computer programme development for the design of IFAS waste water treatment processes, M.S.Thesis, Virginia Tech, Blacksburgs, VA 24061, pp 1540–1549
13. Gebara F (1998) Activated sludge biofilm waste water treatment system, Water Res 33(1):230–238
14. Fadel M, Ibrahim A, Ayoub G (2002) Aerobic hybrid growth process modeling: a parametric sensitivity analysis. In: Proceedings of international symposium on environmental pollution control and waste management, pp 333–348
15. Fouad M, Bhargava R (2005) A simplified model for the steady-state biofilm – activated sludge reactor. J Environ Manage 74:245-253
16. Henze M, Gujer W, Mino T, van Loosdrecht M, Marais, GVR Wentzel MC, Matsuo T (1999) Activated sludge model No. 2D, ASM 2D. Water Sci Technol 39(1):165–182
17. Boltz JP, Johnson BR, Daigger GT, Sandino J (2009) Modeling integrated fixed-film activated sludge and moving- bed biofilm reactor systems I: mathematical treatment and model development. Water Environ Res 81(6):555–575
18. Wanner O, Morgenroth E (2006) Biofilm modeling with AQUASIM. Water Sci Technol 49(1):137-144, IWA Publishing
19. Li C, Ji M, Li X, Wang M (2011) Municipal wastewater treatment in a new type biocarrier reactor. Procedia Environ Sci 10:962–967
20. Golla PS, Reddy MP, Simms MK, Laken TJ (1994) Three years of full scale captor process operation at Moundsville WWTP. Water Sci Technol 29:175181
21. Wanner J, Kucman K, Grau P (1988) Activated sludge process combined with biofilm cultivation. Water Res 22:207–215
22. Daniel D, Christensso M, Odegaard H (2011) Hybrid/activated sludge/biofilm process for the treatment of municipal waste water. Water Sci Technol 63(6):1121-1129. doi: https://doi.org/10.2166/wst.2011.350
23. Sriwiriyarat T et al (2005) Computer program development for the design of integrated fixed film activated sludge wastewater treatment processes. J Environ Eng 131(11):1540–1549
24. Lowry OH, Rosebrough NJ, Fair AL, Randall RJ (1951) Protein measurement with Folin Phenol Reagent. J Biol Chem 193:265–275
25. Herbert D, Philips PJ Strange JE (1971) Chemical analysis of microbial cell of the book. Meth Microbio Vol.V, Part B, Chapter-3, pp 209–344

Chapter 2
Advantages of Hybrid Bioreactor

The limitations of activated sludge process with regard to surge loading, maintaining uniform biomass concentration, poor settleability in secondary clarifier, inadequacy to withstand toxic and inhibitory substances, etc. have been focused. Thereafter, the efforts have been made to highlight on upgradation of activated sludge process (ASP) with different possible forms. Advantages of attached-growth process and development of aerobic fixed-bed hybrid bioreactor are also highlighted in this section. Chronological developments of various mathematical models with their comparative features and validation with experimental data are also presented. Treatability study of municipal wastewater under both ASP and aerobic fixed-bed hybrid bioreactor is also explored. Finally, critical assessment is done based on evolution of aerobic fixed-bed hybrid bioreactor, its present and future perspective, mathematical modeling of hybrid bioreactor and its validation in the treatment of synthetic and municipal wastewater.

2.1 Limitations of Activated Sludge Process

Activated sludge process has experienced several operational problems since its inception. The major problems encountered by this process are depicted below.

2.1.1 Substantial Volume of the Tank

Maintaining uniform biomass concentration in the aeration tank is a great challenge in the ASP, requiring for sludge recirculation. In order to accommodate the recirculated flow, additional volume of the aeration tank needs to be provided. As a result, the overall volume of the aeration unit gets increased leading to higher capital cost.

S. Sarkar and D. Mazumder, *Fixed Bed Hybrid Bioreactor*, Green Energy and Technology,
https://doi.org/10.1007/978-981-33-4546-1_2

2.1.2 *Surge Loading*

Sudden increase in the organic load of influent wastewater causes a shock to the microorganisms, which are exposed to a certain organic concentration in the aeration unit. In that case, inhibition may also take place due to additional substrate (organic matter) leading to process failure.

2.1.3 *Inability to Maintain Uniform Biomass Concentration*

The suspended biomass in the aeration tank is subjected to wash out when there is no recirculation of biomass. This problem is most acute for low-strength wastewater. At the same time, concentration of the settled sludge in secondary clarifier varies and therefore the rate of biomass recirculation also needs to be adjusted. However, continual adjustment is not always feasible, causing a fluctuation in the biomass concentration of the aeration tank.

2.1.4 *Poor Settleability in Secondary Clarifier*

There is a limitation on clarifier solids loading that put an upper limit on the amount of biomass in the suspended-growth system. If MLSS content is too high, especially greater than 3500 mg/l the sludge matrix becomes tight and poorly settles in the secondary clarifier. As a result, the quality of effluent also deteriorates due to improper solid–liquid separation. This can also cause the dissolved oxygen (DO) content to drop, which affects the aerobic degradation of organic matters. Two principal types of enhanced biomass, i.e., sludge bulking problems like filamentous bulking and viscous bulking are observed in ASP process. Filamentous bulking is due to the growth of filamentous organisms, whereas viscous bulking is on account of the excessive amount of extracellular polymers producing the sludge with a slimy, jelly-like consistency [1].

2.1.5 *Inadequacy to Withstand Toxic and Inhibitory Substances*

Slowly biodegradable substances in the wastewater are generally the aliphatic and aromatic organic compounds (e.g., Phenol, Benzene, Toluene, Xnylene, Chloroform, etc.), and their salts with sulfate, chloride, etc. which pose unique problems in wastewater treatment, due to their resistance to biodegradation and potential toxicity to the environment and human health. These compounds are difficult to treat in conventional activated sludge process and are designated as refractory or recalcitrant. Presence of recalcitrant compounds and high concentration of total dissolved inorganic solids associated with low BOD/COD ratio (<0.35) of wastewater is generally considered as complex or slowly biodegradable. There is always inhibitory effect of those substances in biological treatment of wastewater.

2.1.6 High Energy Requirement During Operation

In traditional ASP, biomass is maintained through extended aeration system with necessary recirculation ratio, which increases the pumping cost. As the biomass are susceptible to wash out, concentrated return activated sludge needs to be recirculated to the aeration tank, resulting in high power input toward pumping.

2.2 Efforts Made on Upgradation of Activated Sludge Process

Integrated film activated sludge process was first developed in the USA in 1980, where biofilm was inserted into the aeration tank [2, 3]. Hybrid bioreactor is mainly of two types: "dispersed media," i.e., movable type biofilm entrapped in the aeration tank and "fixed media," i.e., sheet media or fabric media fixed in place in the aeration tank. The later one is said to be IFAS technology, which provides for additional biomass within a wastewater treatment facility in order to meet more stringent effluent discharge standard or increased organic loading without the direct need for additional tank [4].

Dispersed biofilm media in case of hybrid bioreactor with movable biofilm are suspended by the flow induced by the aeration system. Fine bubble diffusers are used for aeration in case of fixed film hybrid bioreactor and coarse ones are used in case of movable film hybrid bioreactor.

The technology of hybrid bioreactor system can be applied as a retrofitting of the existing ASP system without expanding the volume of the reactor. Hybrid bioreactor technology has been incorporated into municipal and industrial wastewater facilities for upgrading or retrofitting plants in many variations of suspended-growth systems. In case of upgraded plants, additional treatment capacity can be obtained without increasing the volume of the existing tank. The secondary treatment process was upgraded to integrated fixed-film activated sludge process by adding the media into its existing basin in Broomfield wastewater treatment plant for improving the biological nutrient removal process [5]. A municipal ASP treating domestic wastewater was upgraded by the incorporation of synthetic looped cord media supporting fixed biofilm growth at Harbor road WWTP, Florida. The plant was upgraded to IFAS system to meet the effluent quality in terms of unionized ammonia of 0.02 mg/l for river discharge (Zammataro et al. 2004). Biological nutrient (ammonium nitrogen) removal was performed by hybrid bioreactor introducing porous ceramic particles into the reactor and the biofilm thickness was observed using scanning electron microscope [6]. The treatment outcomes revealed that the attached biomass has a stronger ability to resist the nitrification inhibitor. The marble particles, as a suspended carrier was added in the activated sludge process for promoting its nitrification process [7].

Sponge-type media was also used as biofilm media in IFAS system [8]. Dispersed plastic foam particles, partly submerged rotating disk and fully submerged plastic disks were also used as a movable media for attaching the biomass in the hybrid bioreactor [9]. A moving bed biofilm pretreatment was integrated into the plant with improved sludge separation in the secondary clarifier for attaining the limit of COD discharge into nearby river [10, 11]. Activated sludge biofilm wastewater treatment system was developed as a hybrid modification of existing ASP by inserting plastic nets vertically into the aeration tank [12–14]. Hybrid bioreactor was applied for biological nutrient removal in industrial wastewater treatment [6]. Porous ceramic particles were introduced into the reactor to provide the surface for biomass attachment. It was observed that attached biomass is superior to suspended biomass for nitrifying wastewater containing toxic organic compounds.

Rapid nitrification was achieved by employing macro-porous cellulose carrier for attachment of nitrifying bacteria [15]. Nitrification was also done using chalk as a biomass carrier and for attaching nitrifying biomass in a hybrid bioreactor [16]. A similar line of research using talqueous powder as a medium of attachment surface for improving the performance of existing activated sludge plant without changing its infrastructure was also carried out [17]. 30% increase in nitrification with improvement in sludge volume index (SVI) compared to the existing ASP was observed during this study. The said research was also done using the ferro-magnetic powder in the similar type of reactor [18]. Fibrous carriers were used in anoxic zone in an aeration tank for enhancing the removal of nitrogen and phosphorus [19].

2.2.1 The LINPOR Process

This modified activated sludge system utilized a suspended media of highly porous plastic foams cubes as a movable biofilm media which occupied 10–30% of the liquid volume in an aeration tank. The process produces a treated effluent quality far better than the conventional ASP process. The LINPOR system can be configured according to the level of treatment required (carbonaceous, ammonia reduction or total nitrogen reduction) and the process is ideally suited for upgrading existing systems with minimum modifications to the existing facilities [20, 21] (Fig. 2.1).

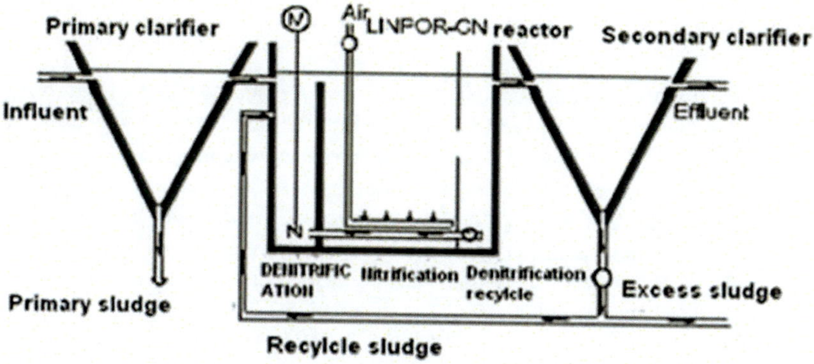

Fig. 2.1 Schematic diagram of typical LINPOR-CN system (*Adapted from* [21])

2.2.2 *The Suspended Carrier Biofilm Process (SCBP)*

It is also known as Kaldnes moving bed biofilm reactor, which was applied as extension of existing treatment facilities [22–25]. Here, the dispersed biofilm media was high-density polyethylene for supporting the biofilm growth. The increased biomass concentration in the reactor, and much higher solid retention time (SRT) of the attached biomass result in efficient carbon removal and improved nitrification and denitrification, through process intensification [26]. Compared to a conventional activated sludge process, a similar degree of treatment or effluent standard can be achieved in a much smaller volume [22]. The schematic diagram of suspended carrier biofilm process is shown in Fig. 2.2.

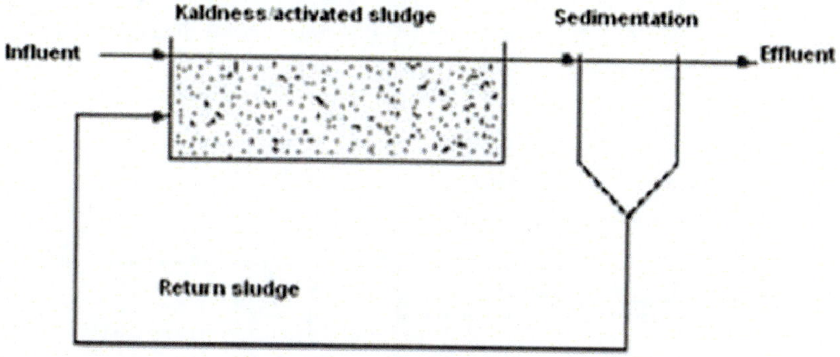

Fig. 2.2 SCBP in form of Kaldnes moving bed system accommodated in the same ASP reactor

2.2.3 *Integrated Fixed-Film Activated Sludge (IFAS) System*

Work on this system was first started in the 80 s and 90 s in the USA. IFAS system adds the benefits of fixed film systems into the suspended-growth activated sludge process. ASP has process flexibility and provides a high degree of treatment, whereas fixed-film processes are inherently stable and resistant to organic and hydraulic shock loading [27, 28]. The capacity of existing ASP system can be enhanced by IFAS process without additional tankage. Two types of media like "dispersed media" and "fixed media" were used in this system. The materials used for the attachment surface are high-density polyethylene. The schematic diagram of IFAS process is shown in Fig. 2.3.

2.3 Advantage of Biofilm Process

Biofilms are increasingly important in biological treatment of wastewater like attached-growth process because of some inherent advantages including low energy consumption, easy maintenance, better stability and excellent biomass retention. Immobilized biofilms are better alternatives to the traditional suspended-growth bioreactor systems because biofilm process can be easily operated under continuous flow mode with minimal biomass loss. Biofilm plays an important role in biological treatment of wastewater as high biomass concentration allows large volumetric loadings and good effluent quality requiring for low energy and these can be maintained without the need for sludge separation and sludge recycling [29]. There is also no contamination at the inner depth of a biofilm under toxic environment making it applicable for treating refractory substances. Moreover, the biofilm process exhibits improved physical and chemical stability of the biocatalyst along with low power requirement [30]. In a biofilm reactor, the biofilm being attached to immobile solid media utilizes substrate as the liquid passes through the reactor in a quasi-plug-flow manner. Biomass makes it applicable for treating wastewater containing refractory

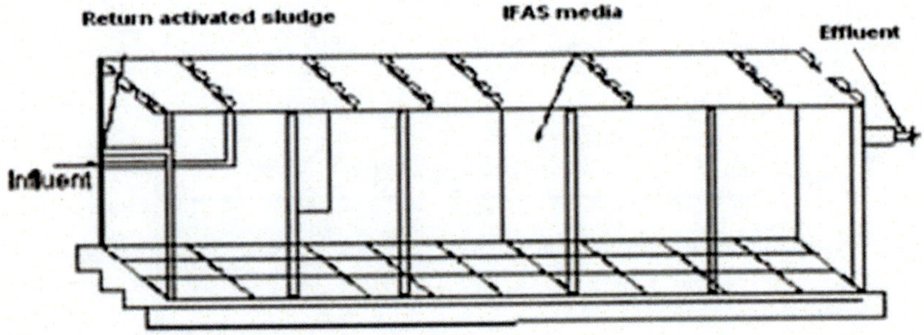

Fig. 2.3 Schematic diagram of a typical IFAS system

substances also, without any trace of contamination into it. The physical and chemical stability of the biocatalyst can be improved by biofilm process with low power input [30].

2.4 Aerobic Fixed-Bed Hybrid Bioreactor

Aerobic Hybrid bioreactor is the integrated biofilm activated sludge system, where both suspended and attached-growth of biomass occur simultaneously. In fact, this system adds the benefits of biofilm system into the suspended-growth activated sludge process by inserting the biofilm surface into the aeration tank for cultivating the biomass. Generally aerobic hybrid bioreactor is completely mixed system, where proper mixing is provided for biomass suspension, substrate transfer and oxygen diffusion. In the last few decades, hybrid bioreactor comprising of suspended- and attached-growth biomass has been adopted widely for the treatment of wastewater [24, 31, 32]. Integrated fixed-film activated sludge process was first developed in the USA in 1980 where biofilm was inserted into the aeration tank [2, 3, 17].

2.4.1 Performance of Fixed-Bed Aerobic Hybrid Bioreactor

It was observed that the integrated-fixed film activated sludge (IFAS) system could increase the effective MLSS in an aeration basin as much as 3000 mg/l. Rostron et al. [33] conducted performance studies with different fixed film media under varying input conditions. Combined activated sludge biofilm process was used for studying the effect of solid retention time (SRT) and dissolved oxygen (DO) on chemical oxygen demand (COD) [34]. It was observed that for efficient nitrogen removal, SRT should be longer than 10 days. Later on, activated sludge biofilm system was developed as a hybrid modification of existing ASP by inserting plastic nets vertically into the aeration tank [12–14], [35]. Hybrid bioreactor was also applied for biological nutrient removal in industrial wastewater treatment [6, 36]. It was noticed that attached biomass is superior to suspended-biomass for nitrifying wastewater containing toxic organic compounds [37]. Rapid nitrification was performed by employing macro-porous cellulose carrier for attachment of nitrifying bacteria [15]. Nitrification was also accomplished using chalk as a biomass carrier and for attaching nitrifying biomass in a hybrid bioreactor [16]. Fibrous carriers were used in anoxic zone in an aeration tank for enhancing the removal of nitrogen and phosphorus [19]. Rutt et al. [5] reported one such case study on integrated fixed-film activated sludge (IFAS) at Broomfield Coke oven Wastewater Treatment Plant (WWTP). A laboratory-scale shaft-type aerobic hybrid bioreactor was investigated to check the extent of removal of carbonaceous organic matter [38]. About 80% COD removal was achieved at a HRT 8 h for an initial COD concentration between 1000 and 3500 mg/l. Hybrid bioreactor was used for studying the nutrients removal from municipal wastewater [39, 40]. It was found that the hybrid bioreactor ensured a better removal of nutrient than the purely attached biofilm system [41, 42]. Thalla et al. [43] conducted laboratory experiments on activated sludge biofilm system under the influence of a composite medium and using vinasse as the substrate. It was noticed that concentration of biomass attached in the media was higher than that in suspension. The biomass around the media played a significant role in the overall process. Municipal wastewater containing oils and fats, dairy wastewater, meat industry wastewater, hospital wastewater, rapeseed oil industry wastewater were also treated with hybrid bioreactor [18, 39, 44–49].

An aerobic hybrid bioreactor was operated for 120 days at Tilanjin, China [39] for COD and nutrients removal from municipal wastewater. The removal efficiency of COD and ammonium nitrogen was found 76% and 72.6%, respectively.

2.4.2 *Integrated Fixed-Film Activated Sludge (IFAS) System—At a Glance*

In the early 80 s, work started in the USA on integrated fixed-film activated sludge (IFAS) system. IFAS systems add the benefits of fixed-film systems into the activated sludge process. The additional biomass provided by introducing fixed-film media directly into the suspended-growth reactor does not increase clarifier solid loading that often limits the treatment capacity of existing ASP system. IFAS technology fulfills the need for enhancing activated sludge plant capacity, with little or no additional tankage, by virtue of the additional fixed biomass. In IFAS system, a variety of media such as structured sheet media, knitted fabric media, etc. can be used to introduce the biomass in the activated sludge basin. Nowadays, the most common material of fabrication for media is either high-density polyethylene (HDPE) or polypropylene (HDPP) or polyurethane, which is chemically inert and can be molded into any shape. The geometric shape and size of the media directly influence the specific surface area and the cost of production. To make the IFAS system successful, it is essential that the internal pores remain open to maximize the effective surface area [27, 28]. Biomass populations in IFAS aeration basins are not susceptible to biomass washouts during hydraulic surges as they are fixed in place and the fixed biomass acts as a "re-seeder" to get the system returned to normal operation quickly after the same.

2.5 Mathematical Modeling of Aerobic Fixed-Bed Hybrid Bioreactor

In order to design a hybrid biological process, a rational and simplified model is very much useful. A good model not only elucidates a better understanding of the complicated biological fundamentals, but is also essential for process design [50–52], process start-up [53], dynamic predictions [54–56], process control [57] and process optimization [58]. Hybrid bioreactor model is essentially based on integration of both the suspended- and attached-growth kinetics, where two types of microorganism act simultaneously. There are limited number of methods available for solving problems on hybrid bioreactor system. This is because the biological reactions involving the conversion of the substrates by both suspended- and attached-growth biomass in a concurrent way are highly complex. In general, this model is assumed to be consisting of traditional suspended growth and biofilm models out of which solution of latter one is much tedious and cumbersome. Although hybrid bioreactor exhibiting both suspended and attached growth is widely used for biological treatment of wastewater, the reliable analysis for predictions pertinent to the performance of the said reactor is still unknown. Mostly the kinetic models assuming single growth, i.e., either suspended or attached were derived. Even where the competition for rate-limiting substrates between two growths (both suspended and attached) were considered, no unique accurate solution was obtained from the steady-state substrate and biomass balance equations Reliable analysis for the appropriate process design leading to predictions pertinent to the performance of the said reactor is really a challenge to the researchers.

2.5.1 *Concept of Modeling of Hybrid Bioreactor*

In this reactor, a fraction of substrate is used by suspended biomass and the remaining by attached biomass resulting in competition between the two growths for the substrate.

The schematic diagram of a typical hybrid bioreactor with fixed-bed biofilm is shown in Fig. 2.4, where attached-growth biomass prevails over some media and there is a steady suspended biomass concentration. The biomass separation takes place in the secondary clarifier and the effluent leaves out from the reactor by overflow mode.

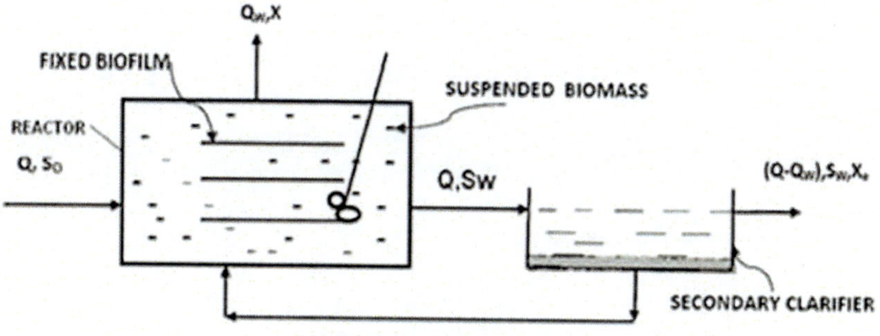

Fig. 2.4 Conceptual diagram of a typical aerobic hybrid bioreactor

The mathematical model of a hybrid bioreactor is an integration of the biofilm model and the classical suspended-growth model. Out of these two models, suspended-growth model is almost unique and simple, which is based on Monod's growth kinetics. But, the analysis of Biofilm model appears to be cumbersome on account of its complexity [59, 60]. Besides, the characteristic of biofilm uptake is quite different from the suspended-growth biomass [61]. Although there are many existing models on biofilm, the procedure of solution is very tedious and the results of outputs are found approximate. There is a limited number of models in the integrated biofilm and suspended-growth process.

2.5.2 *Mathematical Modeling of Suspended-Growth Process*

In 1982, a task group was formed by International Association on Water Pollution Research and Control (IAWPRC) to facilitate the application of models for designing and operating activated sludge systems. Many models have since been developed to describe various complex activated sludge systems. Grady [62] initially presented a single sludge system model, which consisted of seven processes including carbon oxidation, nitrification, and denitrification, which was the outline for a model, agreed upon by the task group. The single sludge system model was developed to describe a single sludge system with the matrix approach proposed by IAWQ (Bidstrup 1988). A preliminary model was evaluated by Dold and Marais (1986) and this led to the development of Activated Sludge Model No. 1 (ASM1). Modifications to that model were made and adopted as a final version of ASM1 [62].

Model Equations for ASM1

The rate of change of $X_{B,H}$ is given by.

$$\begin{aligned}\frac{dX_{B,H}}{dt} &= \mu_H \frac{S_s}{K_s + S_s} \frac{S_O}{K_{O,H} + S_O} X_{B,H}. \\ &\quad + \eta_g \mu_H \frac{S_s}{K_s + S_s} \frac{K_{O,H}}{K_{O,H} + S_O} \frac{S_{NO}}{K_{NO} + S_{NO}} X_{B,H} - b_H X_{B,H}\end{aligned} \tag{2.1}$$

The rate of change of $X_{B,A}$ is given by

$$\frac{dX_{B,A}}{dt} = \mu_A \frac{S_{NH}}{K_{NH} + S_{NH}} \frac{S_O}{K_{O,A} + S_O} X_{B,A} - b_A X_{B,A} \tag{2.2}$$

The rate of change of S_s is given by

$$\frac{dS_s}{dt} = \left[-\frac{\mu_H}{Y_H} \left(\frac{S_s}{K_s + S_s} \right) \left\{ \left(\frac{S_O}{K_{O,H} + S_O} \right) + \eta_g \left(\frac{K_{O,H}}{K_{O,H} + S_O} \right) \left(\frac{S_{NO}}{K_{NO} + S_{NO}} \right) \right\} \right.$$

$$+ k_h \frac{X_s/X_{B,H}}{K_X + (X_S/X_{B,H})} \left\{ \left(\frac{S_O}{K_{OH} + S_O} \right) + \eta_h \left(\frac{K_{O,H}}{K_{O,H} + S_O} \right) \left(\frac{S_{NO}}{K_{NO} + S_{NO}} \right) \right\} \Bigg] X_{B,H} \tag{2.3}$$

The rate of change of X_s is given by

$$\frac{dX_S}{dt} = (1 - f_p)(b_H X_{B,H} + b_A X_{B,A}) - k_h \frac{X_s/X_{B,H}}{K_X + (X_s/X_{B,H})} \left\{ \left(\frac{S_O}{K_{O,H} + S_O} \right) + \eta_h \left(\frac{K_{O,H}}{K_{O,H} + S_O} \right) \left(\frac{S_{NO}}{K_{NO} + S_{NO}} \right) \right\} X_{B,H} \tag{2.4}$$

The rate of change of X_p is given by

$$\frac{dXp}{dt} = f_p \left(b_H X_{B,H} + b_A X_{B,A} \right) \tag{2.5}$$

The rate of change of X_{ND} is given by

$$\frac{dX_{ND}}{dt} = (i_{XB} - f_p i_{XP})(b_H X_{B,H} + b_A X_{B,A}) - k_h \frac{X_{ND}/X_{B,H}}{K_X + (X_s/X_{B,H})} \left\{ \left(\frac{S_O}{K_{O,H} + S_O} \right) + \eta_h \left(\frac{K_{O,H}}{K_{O,H} + S_O} \right) \left(\frac{S_{NO}}{K_{NO} + S_{NO}} \right) \right\} X_{B,H} \tag{2.6}$$

The rate of change of S_{ND} is given by

$$\frac{dS_{ND}}{dt} = (-k_a S_{ND} + k_h \frac{X_{ND}/X_{B,H}}{K_X + (X_s/X_{B,H})} \left\{ \left(\frac{S_O}{K_{O,H} + S_O} \right) + \eta_h \left(\frac{K_{O,H}}{K_{O,H} + S_O} \right) \left(\frac{S_{NO}}{K_{NO} + S_{NO}} \right) \right\} X_{B,H} \tag{2.7}$$

The rate of change of S_{NH} is given by

$$\frac{dS_{NH}}{dt} = [-i_{XB} \mu_H \left(\frac{S_s}{K_s + S_s} \right) \left\{ \left(\frac{S_O}{K_{O,H} + S_O} \right) + \eta_g \left(\frac{K_{O,H}}{K_{O,H} + S_O} \right) \left(\frac{S_{NO}}{K_{NO} + S_{NO}} \right) \right\} + k_a S_{ND}] X_{B,H} - \mu_A \left(i_{XB} + \frac{1}{Y_A} \right) \left(\frac{S_{NH}}{K_{NH} + S_{NH}} \right) \left(\frac{S_O}{K_{O,A} + S_O} \right) X_{B,A} \tag{2.8}$$

The rate of change of S_{NO} is given by

$$\frac{dS_{NO}}{dt} = -\mu_H \eta_g \left(\frac{1 - Y_H}{2.86 Y_H} \right) \left(\frac{S_S}{K_S + S_S} \right) \left(\frac{K_{O,H}}{K_{O,H} + S_O} \right) \left(\frac{S_{NO}}{K_{NO} + S_{NO}} \right) X_{B,H} + \frac{\mu_A}{Y_A} \left(\frac{S_{NH}}{K_{NH} + S_{NH}} \right) \left(\frac{S_O}{K_{O,A} + S_O} \right) X_{B,A} \tag{2.9}$$

The rate of change of S_O is given by

$$\frac{dS_O}{dt} = -\mu_H\left(\frac{1-Y_H}{Y_H}\right)\left(\frac{S_S}{K_S+S_S}\right)\left(\frac{S_O}{K_{O,H}+S_O}\right)X_{B,H} - \mu_A\left(\frac{4.57-Y_A}{Y_A}\right)\left(\frac{S_{NH}}{K_{NH}+S_{NH}}\right)\left(\frac{S_O}{K_{O,A}+S_O}\right)X_{B,A} \quad (2.10)$$

The nomenclature of all the above parameters is listed below.

IAWQ model parameters	Symbol
Heterotrophic yield	Y_H
Autotrophic yield	Y_A
Mass N/mass COD in biomass	i_{XB}
Mass N/mass COD in products from biomass	i_{XP}
Heterotrophic max. specific growth rate	$\hat{\mu}_H$
Heterotrophic decay rate	b_H
Half-saturation coefficient (hsc) for heterotrophs	K_S
Oxygen hsc for heterotrophs	$K_{O,H}$
Nitrate hsc for denitrifying heterotrophs	K_{NO}
Autotrophic max. specific growth rate	$\hat{\mu}_A$
Autotrophic decay rate	b_A
Oxygen hsc for autotrophs	$K_{O,A}$
Ammonia hsc for autotrophs	K_{NH}
Correction factor for anoxic growth of heterotrophs	η_g
Ammonification rate	k_a
Max. specific hydrolysis rate	k_h
Hsc for hydrolysis of slowly biodeg. substrate	K_X
Correction factor for anoxic hydrolysis	H_h

The Activated Sludge Model No. 1 (ASM1) was a useful tool for research, development and optimization of biological nitrogen removal processes. Subsequently, many other models have been developed building upon the ASM1 model. However, ASM1 does not include the phenomenon of excess biological phosphorus removal (EBPR). Extension of ASM1 to include EBPR was proposed and developed by IAWQ and called ASM2. ASM2 includes both the excess biological phosphorus removal and nitrogen removal (Henze et al. 1994). Barker and Dold [63] developed their general model, which includes 36 processes, among them, the EBPR is a component. The anoxic growth of poly-P microorganisms also has been incorporated in the model. Anoxic and anaerobic hydrolysis of slowly biodegradable COD is incorporated into this model as well as the fermentation of readily biodegradable organic materials. The general model is based on the original ASM1 and the Wentzel Model (1989). Thus, the Activated Sludge Model No. 2 (ASM2) presents a concept for dynamic simulation of combined biological processes for chemical oxygen demand (COD), nitrogen and phosphorus removal.

Henze [64] proposed ASM No. 2d that includes phosphorus removal by poly-P microorganisms during denitrification. In ASM2d model, denitrifying phosphorus-accumulating organisms (PAOs) were included. Thus ASM2d should be used as a basis for modeling of simultaneous biological phosphorus uptake and nitrification–denitrification. Gujer (1999) presented an ASM No. 3 model, which is a modification of ASM No. 1. A substrate storage process was introduced into this version as a new process and it is assumed that there is no COD flow from nitrifier decay. The modeling of heterotroph and autotroph decay processes was clearly separated in ASM No. 3. ASM3 models provide a common base for the simulation of nitrogen removing activated sludge systems for chemical oxygen demand as well as organic carbon-based characterization of wastewater and biomass.

2.5.3 Mathematical Modeling of Attached-Growth Process

Various researchers have developed their steady-state biofilm models on biological attached-growth process in wastewater treatment which have certain limitations [65, 66]. Runge–Kutta finite difference technique was applied in approximate solution of second-order differential equation of mass balance of substrate in the biofilm [67]. The above biofilm model had a drawback that it had not considered any substrate balance in attached growth including the term "specific surface area," "hydraulic retention time," etc. which are the crucial parameters in the design of hybrid bioreactor. It is a real fact that, for calculation of the thickness of stagnant liquid layer both specific surface area (a) and hydraulic retention time (θ) are required. Moreover, to maintain the steady-state biofilm model it is necessary to check whether the value of effluent substrate concentration is greater than minimum substrate concentration ($S_{\min}$), required to maintain the biofilm thickness or not. No such consideration was made in the biofilm model by Williamson and Mccarty [67]. Apart from that, in the biofilm model by Williamson and Mccarty [67] nomographs were used, so the output results might be approximate and susceptible to human error.

Later on, using a new concept on steady-state biofilm model, the total biofilm thickness was measured from the mass balance on active biomass inside the biofilm [68, 69]. However, the effective thickness of biofilm could not be evaluated using this model. Moreover, instead of a simplified analytical solution to solve second-order differential equation of mass balance of substrate in the biofilm, a pseudo-analytical approach was used, which was approximate and cumbersome also. Pseudo-analytical solution with various dimensionless parameters appeared to be very complicated and time-consuming too. In this method, two levels of iteration were required to find out the value of effluent substrate concentration.

Eventually, the differential equation pertaining to biomass balance was eliminated by semi-empirical algebraic expression in order to develop a single substrate dimensionless biofilm model. The said model was solved numerically to generate a nomograph that correlates dimensionless substrate concentration at biofilm–liquid interface, dimensionless substrate flux, dimensionless biofilm thickness and dimensionless minimum substrate concentration that is needed to sustain the flux [70]. The solution of this model was also approximate one and only the total biofilm thickness could be determined. Another biofilm model was developed using normalized loading curves with dimensionless variables [71]. The model derived an approximate solution, which could not determine the effective biofilm thickness. There was also a chance of human error while studying the output variables from the normalized loading curves. Normalized loading curve is given for a particular value of S_{min*} (i.e., S_{min}/K); for any other value of S^* interpolation needs to be done between the values of two curves yielding an approximate output result. Moreover, normalized loading curve can be used only if effluent substrate concentration is known, i.e., targeted to a certain value. By normalized loading curve, the substrate flux (J) can only be evaluated. Saez et al. [72] proposed an accurate pseudo-analytical solution for steady state biofilm model. However, it has also not considered any substrate balance in attached growth including the term specific surface area, hydraulic retention time, etc., which are the crucial variables in designing the biofilm reactor. The effective thickness of biofilm could also not be calculated using this model.

Biomass loss from the attachment surface is a critical issue in a biofilm reactor. The biomass loss rate from the attachment surface for a steady-state biofilm reactor was evaluated by the following equation (Lee et al. 1999).

$$b_s = \frac{QX}{X_f L_f (S_a V)} \tag{2.11}$$

where, Q = flowrate, X = effluent biomass concentration, X_f = biofilm density, L_f = biofilm thickness, S_a = specific surface area of the media attachment, V = empty-bed reactor volume. However, there was no analytical procedure found out in earlier research for determining b_s in case of hybrid bioreactor. Apart from that, one steady-state biofilm model was proposed considering the biofilm attached to substratum [73]. But, no dislodging effect, i.e., no shear loss of biofilm from the attachment surface was considered in that steady-state one-dimensional biofilm model.

One mathematical model for a fixed biofilm reactor treating phenolic wastewater was developed using Fick's diffusion law and Haldens' Kinetics. The growth of biofilm, mass balance equation of substrate and suspended biomass were expressed in form of three differential equations. The model was analytically solved by combination of collocation method and Gear's method. By orthogonal collocation method, the partial differential equations in dimensionless form were ultimately converted to nine ordinary differential equations, which were solved by using Gear's method [74].

Apart from that, an weighted average of the analytical solutions for zero and first-order basic diffusion reactions was also developed to demonstrate the flux of substrate through the biofilm [75, 76]. The objective of this model was to provide a simple tool to determine substrate flux and effluent substrate concentration in steady-state biofilm system. The model was based on the principle that difference between the mass of the substrate between the influent and effluent is transferred toward the biofilm by diffusion due to substrate gradient in x direction and then it is converted by the biomass present in the biofilm. Consequently, a steady-state biofilm model for the simultaneous utilization of dual substrates like an electron donor and an electron acceptor was developed, where the flux of the electron donor is balanced by the flux of electron acceptor (Shaoying et al. 2005). The model was based on substrate transfer equation through the diffusion layer including the biofilm and the reduction–oxidation reaction of two substrates catalyzed biologically. The dual substrate model produces lower flux compared to single substrate one for the limiting substrate in a deep biofilm.

A model of a fixed-bed biofilm reactor treating pharmaceutical wastewater was developed using Nelder-Mead optimization method in conjunction with first-order kinetics of biofilm process. The model was intended for accurate estimation of kinetic and biofilm thickness parameters ensuring reliable predictive performance of the biofilm reactor model [77]. Another simplified model for the fixed biofilm reactor has been recently developed to formulate the relationship between the substrate concentrations at both entry and exit, at the biofilm liquid interface and at the biofilm attached surface along with average substrate flux in the biofilm, substrate flux at the biofilm liquid interface and effective biofilm thickness [78].

2.5.4 *Mathematical Modeling of Aerobic Hybrid Bioreactor System*

In the earlier research, one steady-state substrate balance for both suspended and attached growth and the biomass balance for the suspended growth were used to develop a model for the hybrid bioreactor considering both attached and suspended growth simultaneously [79]. One computer program was developed for integrated fixed-film activated sludge system for removing soluble COD and nutrients [80]. Later hybrid bioreactor model was applied in a typical activated sludge process, where biofilm was provided with plastic nets vertically inside the tank [12, 81]. In order to analyze an aerobic hybrid process, a simplified mathematical model was developed for a steady state biofilm activated sludge reactor to calculate the substrate flux in the biofilm under substrate limiting condition [13]. Besides this, activated sludge biofilm model was developed under varying dissolved oxygen for the removal of substrates and nutrients by a combined anaerobic /anoxic/oxic rotating biological contactor. In this model, existing mass balance equation within a suspended activated sludge system was modified by incorporating the substrate flux resulted from biofilm especially in the oxic tank. This modified equations depicting the component transformation in biofilm formed a partial differential equation system. The set of equations was solved by the method of lines and the entire model was implemented in a computer program "GMTNCU1.EXE" coded with FORTRAN language [82].

A mechanistic and empirical IFAS model package was developed for predicting carbonaceous removal and nitrification at steady-state condition for activated sludge system with integrated sponge media in aerobic zone [83]. The respective computer program also predicted the amount of sponge media needed for optimizing the design of IFAS system. The drawback of the model is that it is only applicable for the sponge media. The said IFAS system consisted of two anaerobic, two anoxic and three aerobic reactors in series with sponge media installed in aerobic zones running in parallel with a control system without having any media.

The computer program solved all mass balance equations of both substrate and biomass simultaneously with a set of non-linear equations using modified Newton–Raphson techniques. Another drawback of this model was that no biofilm yield and sloughing rates were considered.

Effort has also been made to develop a simplified mathematical model for designing the steady-state biofilm activated sludge reactor under limiting substrate condition [13]. In that model, a simplified expression of J (substrate flux) was used for avoiding the approximate graphical solution for obtaining J as developed by Suidan et al. [70]. However, the said model is based on some dimensionless parameters and it can be analyzed only if the effluent substrate concentration is pre- assigned. Earlier one activated sludge model (ASM2d) was developed for biological phosphorus removal with simultaneous nitrification–denitrification in the activated sludge process [64]. The said model was further extended to a steady-state IFAS model by Boltz et al. [84] using the input taken from biofilm modeling techniques [1]. The said IFAS model considered competition between the biofilm and suspended biomass for macronutrients, electron donor and electron acceptor substrates. The theoretical considerations in the said model include simultaneous diffusion and Monod type reaction kinetics inside the biofilm. The drawback of this IFAS model is that the biofilm thickness (L_f) needs to be known as a priori to run the analysis and the fate of soluble COD is poorly understood.

Eventually, a simplified mathematical model was proposed to provide an accurate tool for describing the steady-state suspended sludge—biofilm system in the treatment of municipal wastewater [39]. The model was based on a steady-state condition for biomass and substrate with a rate-limiting substrate concentration in the reactor. Monod kinetics was used to describe the biomass growth, whereas, the transport of the dissolved components in the liquid phase was demonstrated by molecular diffusion as per Fick's second law.

2.5.5 *Validation of Various Modeling of Aerobic Hybrid Bioreactor*

Most of the mathematical models on hybrid bioreactor were firstly illustrated using one or more sets of available data for the purpose of analytical validation. Glimpses of such analytical validation are presented in Table 2.1.

At the same time, a few laboratory-scale studies were also carried out on the hybrid bioreactor developed by various researchers. Consequently, they compared the experimental observations in respect of relevant parameters with those available from the respective models. A comparison of output parameters, viz. S_w, J, L_e and L_f, derived from various models and concerned experimental observations is also presented in Table 2.2.

2.6 Treatability of Municipal Wastewater

2.6.1 *Characteristics of Municipal Wastewater*

The municipal wastewater exhibits more or less uniform characteristics for a particular location. However, the characteristic profile may be changed depending on food habit and lifestyle. There are so many limitations reporting characterization of municipal wastewater prior to their respective treatability study. The glimpses of all such characterization reports are listed in Table 2.3.

2.6.2 *Scope of Treatment of Municipal Wastewater Using Activated Sludge Process and Its Modification*

The most common suspended-growth process used for municipal wastewater treatment is the activated sludge process. The removal of BOD is accomplished by means of microbial uptake under suspended-growth state. This biological method essentially is an aerobic process and takes place in the aeration tank, with the help of oxygen provided through aeration. The microbial growth leads to formation of biomass flocs, which are\settleable in the secondary clarifier. Hence, the clarified effluent is obtained through the effluent line. Domestic wastewater of all types is treated with the help of activated sludge process but mainly for removal of COD, BOD and TSS. BOD and TSS removal efficiencies of the activated sludge plant (Aeration tank + Secondary clarifier) are 91.27% and 86.76%, respectively (Sundara kumar et al. 2010).

For denitrification of municipal wastewater, Modified Ludzack Ettinger (MLE) process was used which comprised anoxic and aerobic tank along with secondary clarifier, and return activated sludge line from secondary clarifier. In this process by

Table 2.1 Brief summary on application of various fixed-bed hybrid bioreactor models

S. No.	Mathematical model developed by	S0 (mg/ cc)	θ (h)	θc (h)	A (cm^{-1})	X_f(mg/cc)	S_w (mg/cc)	J (mg/cm^2/ day)	L_e (cm)	L_f(cm)
1	[79]	0.43	24	12	0.9	25	0.0039	0.2785	**	0.0122
2	[12]	0.16	6.72	300.72	0.154	54.63	0.018	**	**	**
3	[81]	0.32	6.72	307.68	0.616	48.15	0.003	**	**	0.015
4	[13]	0.43	12	6	1.8	25	0.004	0.273	**	0.0123
5	[83]	0.5	74.4	**	**	**	0.025	**	**	**
6	[39]	0.064	16.7	**	**	**	0.000574	**	**	**

Note "**" indicates data are not available

Table 2.2 Results of experimental validation of relevant output parameters (S_w, J, L_e and L_f) for various hybrid bioreactor models

Model developed by	S_0(mg/cc)	θ (h)	A (cm^{-1})	X_f	S_w (mg/cc)		J (mg/cm^2/day)	
[12, 81]	0.16	6.72	0.154	54.63	0.018	0.016	**	**
	0.16	6.72	0.308	54.63	0.01	0.008	**	**
	0.16	6.72	0.462	48.15	0.007	0.006	**	**
	0.16	6.72	0.616	48.15	0.003	0.004	**	**
[13]	0.43	12	1.8	25	0.004	**	0.273	0.27
[83]	0.5	74.4	**	**	0.025	0.025	**	**

Note "**" indicates data are not available

Table 2.3 Characteristics of municipal wastewater from various literatures

S. No.	Component	References		
		[39]	[64]	[85]
1	TS	–	–	350
2	TSS	–	–	100
3	TDS	–	–	250
4	Alkalinity	–	–	50
5	COD	288	260	250
6	BOD	–	–	110
7	NH^4–N	46.5	–	–
8	TKN	47	25	20
9	Total P	9.61	6	4

using readily biodegradable substrate, nitrate associated with return activated sludge was removed to a satisfactory level. Nitrified liquor from aerobic tank was brought back to anoxic tank for nitrification (CPHEEO manual 2010).

The South-Budapest Municipal Wastewater Treatment Plant (SBMWTP) based on the high-load activated sludge process (ASP) was upgraded into nutrient removal in 1998–1999 in Hungary (Melicz 2003). The study conducted over 25 months revealed that besides the efficient pre-denitrification obtained in the AS basin, significant ammonium oxidation occurred in the aerated zone. Sewage treatment plants in Hong Kong used activated sludge process with up to 5 alternate aeration zones (modified Bardenpho process).

Majd Time Cyclic Reactor (MTCR) is a continuous flow suspended-growth process activated sludge system in which all major conversion steps occur in a single tank with four compartments in a time-based cyclic order. Apart from removal of TSS and BOD, it also enhanced the removal of nitrogen, phosphorus as well as ammonia. It received influent into three compartments and effluent only from another compartment, which followed the timing cycle sequence [86].

Activated sludge process including conventional, extended aeration and contact stabilization was applied in pilot-scale study for six months in Isfahan South Municipal WWTP. COD removal efficiency was found between 83 and 92.5%, between 88 and 93.8%, and between 77 and 92% for conventional process, extended aeration and contact stabilization, respectively [87].

2.6.3 *Past experience on treatment of municipal wastewater in aerobic hybrid bioreactor*

The hybrid bioreactor process has significantly improved the efficiency of biological treatment of slowly biodegradable substance and has been widely applied to municipal wastewater treatment in recent years. Hybrid bioreactor technology has been incorporated into municipal wastewater facilities for both new plant installation and upgradation/retrofitting of the overloaded plants in many variations. In case of new plant installation, the reactor volume gets reduced due to additional attached biomass. For the plant upgradation, enhanced treatment capacity can be obtained without increasing the volume of the existing tank. Hybrid bioreactor was used for studying the nutrients removal from municipal wastewater [39]. It was found that the hybrid bioreactor ensured a better removal than the purely attached biofilm system. IFAS process was also installed at Maryland municipal wastewater reclamation facility, where rope-like ringlace IFAS media of 30000 m were installed in a volume of 475 cum. Nitrification in the IFAS section was found more than three times that in the control section without media [24]. A sponge type IFAS media was used at Moundsville wastewater treatment plant, USA in three years of full-scale operation. This is a combined suspended- and attached-growth process where polyurethane foam media was incorporated into ASP process increasing the equivalent MLSS concentration level. The attached and suspended MLSS were maintained between 6500–12,000 mg/l and 900–1700 mg/l, respectively.

The efficiency in removal of BOD, nitrification and denitrification was found 95%, 70–90% and 40–60%, respectively, at a HRT 50–90 min [8].

A study was conducted to see the effect of attached biomass on the filamentous growth of suspended biomass in the nitrification process and found that nitrification became independent of SRT of suspended biomass, since nitrifying bacteria are predominantly attached to the support materials. In this case, dispersed plastic foam particles and fully submersed plastic disks were used as a supporting media for attached biomass [9]. In a cold region, treatment of domestic wastewater was done by a hybrid activated sludge/biofilm process, which contained both suspended biomass and biofilm, usually referred as integrated fixed-film activated sludge (IFAS) process, created by introducing plastic elements as biofilm carrier media into a conventional activated sludge reactor. Pilot-scale study was carried out at the Department of

Hydraulic and Environmental Engineering at the Norwegian University of Science and Technology in Trondheim. The experiment was performed in two phases. The pilot plant was first operated with a constant hydraulic retention time (HRT) of 4.5 h, while in the second phase the influent flow was increased so that HRT became 3.5 h. The average temperature was maintained near 11.5 °C in the overall experimental campaign. The average mixed liquor SRT was maintained 5.7 days. The results revealed that this kind of reactor can efficiently be used for the upgrading of conventional activated sludge plant for achieving a good quality of effluent [88]. Sriwiriyarat et al. [83] also made his research study treating a week municipal wastewater with IFAS using sponge media at different aerobic mean cell residence times for evaluating his model performance with the experimental data and found satisfactory results in removal of carbonaceous organic matter and nitrification.

2.7 Present and Future Prospect of Aerobic Hybrid Bioreactor System

Although there are a few mathematical models on fixed-bed hybrid bioreactor, no attempt has been made till 2014 toward its process design. There is a future scope in research work for exploring this issue and for proposing a mechanism for the process design of fixed-bed hybrid bioreactor. The process design of hybrid bioreactor is necessary to find out its volume, physical dimension and other operational requirements including oxygen for any targeted effluent substrate concentration. Besides this, it may be aimed to calculate the solid retention time (θ_c), average substrate flux into the biofilm (J_{avg}) and effective as well as total biofilm thickness. There must have certain design criteria, which can be checked accordingly for any physical configuration adopted for the reactor.

Determination of kinetic coefficients for the suspended and the attached biomass is extremely required to predict the performance of a fixed-bed hybrid bioreactor. Since these two types of microorganisms are reciprocal to each other, their nature may be assumed almost same. However, there are additional parameters like coefficient of shear loss and decay, which make the hybrid bioreactor model very complex to analyze. The issue of determining the kinetic coefficients was pending over a long time. There is a scope of future research work for exploring this problem and for suggesting a rational mechanism for evaluating the kinetic coefficients in the fixed-bed hybrid bioreactor.

In earlier research, it was observed that effective diffusivities in aerobic biofilm have a significant role to determine whether a section of the biofilm becomes depleted of substrate or dissolved oxygen, thereby rendering that section ineffective, anoxic or anaerobic [89]. Therefore, extent of diffusion through the biofilm needs to be estimated for ascertaining its actual status.

It has been further established earlier that three sets of indicators like biomass partitions (i.e., the proportions of biomass each in suspended and attached growth), substrate removal partitions (i.e., substrate consumed by each of the two growth) and specific substrate utilization rate evaluate the competitiveness among suspended- and attached-growth biomass for a single substrate in an integrated fixed-film activated process [90]. In this regard, specific substrate utilization rate of both the suspended and the attached biomass must be determined to get rid of their individual performance.

Most of the hybrid bioreactor models already developed appeared to be complex and tedious in calculation requiring significant time for analysis. In this context, the simplified model for hybrid bioreactor may be conceptualized in future research work for predicting the relevant outputs rationally. The future scope in innovative simplified modeling of fixed-bed hybrid bioreactor also lies in extending the same for different kinds of substrate like carbonaceous, nitrogenous and phosphorus by incorporating growth kinetics of respective microorganisms. The influence of various process parameters like pH, dissolved oxygen (DO), etc. may also be catered by introducing all the concerned kinetic expressions.

Process design of the fixed-bed hybrid bioreactor using the rational model can also be optimized to make it cost-effective. In that case, configuration of the hybrid bioreactor in the field can be modified to get it accommodated in a small footprint. Another special aspect regarding the hybrid bioreactor system is that innovations of newer attached media with a view to enhance the specific surface area. Apart from that, the kinetic coefficients for different types of wastewater in a hybrid bioreactor system must be determined precisely. Any rational and simplified model can be used judiciously for this purpose so that the prototype design is possible in case of other flow characteristics. Indeed, the design rationales need to be established by operating the hybrid bioreactor under wide fluctuations of the process parameters, before going to such design.

The sensitivity, i.e., influence of any parameter on the overall performance is an important issue also in case of hybrid bioreactor model. The profile of performance variation on account of change in a specific process can be derived, provided the model equations are accurate and dependable. Hence, out of several variables, the most sensitive one can be detected and thereby its optimal range can be ascertained.

2.8 Mathematical Modeling of Aerobic Hybrid Bioreactor System

In most existing models of hybrid bioreactor, the biofilm liquid interface has not been considered properly. Mass transfer at this surface is decisive for the supply of the active biomass with all types of soluble substrates. It is quite reasonable to assume that the biofilm surface is separated from bulk liquid by a transition zone, where the classical biofilm theory can be applied. In several models of biofilm, the thickness of this transition layer was considered as 100 μm (Wanner and Gujer 1986, Suidan et al.

1996). The transition layer is also called diffusion boundary layer as mass transfer in this zone is assumed to be diffusive.

The first model regarding the substrate flux at the biofilm surface through the diffusion process came into light in 1976 [67, 61]. Substrate conversion in the biofilm was assumed to follow Monod kinetics which was coupled to diffusion based on Fick's second law. Thus, a steady-state substrate utilization relationship was proposed as follows.

$$\frac{d^2 S_{\mathrm{f}}}{dz^2} \frac{k X_{\mathrm{f}} S_{\mathrm{f}}}{D_{\mathrm{f}} (K + S_{\mathrm{f}})} \tag{2.12}$$

S_{f} substrate concentration at any point in the biofilm (mg/cm^3),
X_{f} active biomass density within the biofilm (mg/cm^3),
D_{f} molecular diffusion coefficient of the substrate in the biofilm (cm^2/day),
K half-velocity coefficient (mg/cm^3),
z distance into the biofilm (cm),
k maximum specific rate of substrate utilization (per day).

Later on, this diffusion model was used by Rittmann and Mccarty [91] and subsequent researches in the mass balance equation of substrate for the biofilm.

Most of the existing models of aerobic fixed-bed hybrid bioreactor are based on the simultaneous diffusion and Monod-type reaction kinetics of substrate inside the biofilm and separate entity of substrate utilization by the suspended-growth biomass. However, the substrate utilization in case of suspended biomass is also considered to follow Monod kinetics. In most cases, analytical solution was made for one-dimensional biofilm assuming zero substrate gradient at the terminal end of the effective biofilm to evaluate substrate flux across the biofilm surface. In order to determine the substrate concentration at any point of the biofilm, different numerical analysis methods were adopted depending on accuracy desired. Eventually, it has been observed that there is only one method called "*Runge–Kutta method of analysis,*" which can determine the effective biofilm thickness.

Another important issue regarding fixed-bed hybrid bioreactor is that reciprocation of the suspended and the attached biomass. Since, both types of microorganisms prevail in the same reactor system, the products of shear loss and endogenous decay, etc. from the attached biomass contribute to the suspended biomass. Hence, the suspended biomass comprises of mixed microbial population with a variety of species, making the reactor system stable and consistent in performance. It is therefore important to characterize both the suspended and attached biomass to predict about their performance. Moreover, usual washout of the suspended biomass under continuous operation of the fixed-bed hybrid bioreactor can be supplemented by the dislodged fraction of attached biomass. When the rate of dislodgement of attached biomass equals to that of washout, the reactor can be run under uniform total biofilm without any need of recirculation.

In the fixed-bed hybrid bioreactor, there may be shallow to deep biofilm on the attachment surface. When the biofilm is shallow in thickness, oxygen may be available up to the end of biofilm layer making it aerobic as a whole. On the other hand, for the deep biofilm there would be scarcity of oxygen after a certain point of biofilm layer, where anoxic/anaerobic environment will start to develop. This situation is absolutely favorable for denitrification and initiation for phosphorus removal. Hence, there is a great potential of fixed-bed hybrid bioreactor to achieve combined carbon oxidation-nitrification, denitrification and phosphorus removal in the same system depending on the biofilm thickness.

The Monod kinetics was used for developing most of the fixed-bed aerobic, hybrid bioreactor models assuming the substrate is highly biodegradable. However, this kind of reactor system may also be subjected to slowly biodegradable/toxic substances, where the biofilm plays an important role. Under such circumstances, the kinetics should be altered and it may be considered as follows in accordance with Haldane's approach (Gnzen 1999).

$$\mu = \frac{\mu_m S_f}{K_s + S_f + S_f^2/K_i} \tag{2.13}$$

In that case, the procedure of analysis of the hybrid bioreactor system will also change. It is therefore required to establish the kinetics of substrate removal to check out any sort of inhibition.

Although [79] earlier developed in a true sense a hybrid bioreactor model considering both suspended and attached growth simultaneously, no explicit relationship between substrate concentration (*S*) and substrate flux (*J*) was depicted in his model. As a result, it was difficult to determine the flux (*J*) because *S* and *J* are interdependent in his model. Another drawback of this model was that the Regular Falsi technique was applied for the numerical solution of the said model and was found inconvenient and approximate too. Substrate mass balance was written considering simultaneous uptake by both the suspended- and attached-growth microorganisms as

$$S_0 - S_w - \frac{kSw\theta YaJ\frac{bs}{bt}}{(K+Sw)\left(\frac{1}{\theta c}+bd-\frac{YkSw}{K+Sw}\right)} - aJ\theta = 0 \tag{2.14}$$

where, J = substrate flux into the biofilm (mg/cm^2/day), S_0 = substrate on centrations in the bulk liquid (mg/cm^3), S_w = Effluent substrate concentration(mg/cm^3), k = maximum specific rate of substrate use (day^{-1}), θ = empty-bed hydraulic detention time (day), θ_c = solid retention time(day), K = half-velocity coefficient (mg/cm^3), a = specific surface area of supporting media(cm^{-1}). b_s = biomass loss rate due to shearing from biofilm, day^{-1}, b_t = total biomass loss rate from biofilm, day^{-1}, Y = bacteria yield coefficient.

The above equation was developed by adding two individual substrate removal kinetics of suspended- and attached-growth microorganisms. The respective equations are as follows:

$$r(S_0 - S_w) - \frac{pkXS_w\theta}{K + S_w} = 0 \tag{2.15}$$

$$\text{and } (1-r)(S_0-S_w)-aJ\theta = 0 \tag{2.16}$$

(where r = fraction of substrate used by suspended biomass at steady state, p = porosity of hybrid reactor, X = concentration of suspended biomass in hybrid reactor(mg/cm^3).

While developing the overall kinetics of the hybrid bioreactor, Monod's approach was followed assuming no substrate inhibition.

The model developed by Gebara [12] and [81], considered sequential function of attached- and suspended-growth system in series, i.e., first unit was modeled as biofilm reactor and the second unit was modeled as activated sludge one. Obviously, the drawback of the model was that the two growths, i.e., suspended and attached growth of biomass was not taken into account simultaneously. Furthermore, the mathematical model exhibited by [81] is valid for kinetic coefficients k and Y in the range of (2–3) day^{-1} and (0.4–0.45), respectively. The combination of suspended and attached biomass enhanced effluent concentration, solids settling and improved nitrification efficiency.

The model developed by Sriwiriyarat et al. [80] was found very complicated without considering the simultaneous growth of both suspended and attached biomass. Bhargava et al. [13] developed a steady-state biofilm activated sludge reactor to calculate the substrate flux in the biofilm under substrate limiting condition. The limitation of this mathematical model is that the analysis can be made for the targeted effluent concentration, i.e., when the effluent substrate concentration is known. One computer package called "IFAS" considered only sponge-type media, which used as a biofilm attached surface in a hybrid bioreactor [83].

In the model developed by [84], the biofilm thickness Lf needs to be known as a priori to run the analysis and the final fate of soluble COD is not clearly understood.

The literature study clearly depicts that no mathematical model on hybrid bioreactor is accurate one and involving simple steps of calculation. In light of critical issues as stated above, the inherent drawbacks of existing models developed so far have been summarized in Table 2.4. The limitations of each model are also envisaged from the drawbacks of respective model.

Table 2.4 Summary of drawbacks of existing mathematical models on fixed-bed—hybrid bioreactor

S. No.	Mathematical model developed by	Drawbacks
01	[79]	Regular Falsi technique was applied for the numerical solution of the said model and was found inconvenient and approximate too. No model outputs were validated with experimental results. The substrate flux, J calculated by the model is not the average substrate (J_{avg}) through the biofilm. Moreover no effective biofilm thickness (L_e) can be calculated with this model
02	[12]	The suspended and attached growth was not considered simultaneously; instead initially attached growth and subsequently suspended growth was simulated
03	[81]	The suspended and attached growth were not considered simultaneously; instead initially attached growth and subsequently suspended growth was simulated. The mathematical model is said to be valid for k and Y values ranging between (2–3) day^{-1} and between (0.4–0.45) respectively, which indicates its limitation
04	[82]	The calculation of the velocity of biofilm–water interface relative to the biofilm support interface was not shown as indicated in the model. No effective thickness of biofilm could be obtained from this model. The computer program flowchart was not shown although the computer program in Fortran was highlighted
05	[13]	The analysis can be made for the targeted effluent concentration, i.e., when the effluent substrate concentration is pre-assigned
06	[83]	This program supports only sponge-type media used as a biofilm attached surface in a hybrid bioreactor
07	[84]	In this model, the biofilm thickness Lf needs to be known as a priori to run the analysis and the final fate of soluble COD not clearly understood

2.9 Viability of Treatment of Municipal Wastewater Using Aerobic Hybrid Bioreactor

For the municipal wastewater in the medium and small town, the strength of wastewater varies due to disintegrated sewerage systems. For the low strength to medium strength, municipal wastewater main problem for treatment is the sustaining biomass in the reactor. Presently, adequate biomass is maintained through extended aeration and with high recirculation ratio thereby increasing the pumping cost of the treatment. For the low strength, municipal wastewater which is highly biodegradable the suspended biomass is subjected to wash out. Thus, the concept of attached biomass into the suspended-growth system can be reasonably thought for ensuring sufficient biomass in the reactor. The footprint area of the proposed hybrid bioreactor is small compared to the conventional reactor thereby making the economy of the treatment system.

2.10 Validation of Various Models of Hybrid Bioreactor

The mathematical models of the fixed-bed aerobic hybrid bioreactor have already been applied with different sets of input data time to time. The main objective of such application was to demonstrate the mode of operation of respective model. The input data set was chosen as per compatibility of the model under consideration. Indeed, all the models of fixed-bed hybrid bioreactor are not capable to derive all the desired output parameters. A brief summary on application of various aerobic fixed-bed hybrid bioreactor models is already highlighted in Table 2.1. Table 2.1 depicts that the substrate flux (J) could be obtained in two cases, i.e., using the methods prepared by Lee [79] and Fouad and Bhargava [13]. Although the effluent substrate concentration was estimated by all the methods, none of them could determine the effective biofilm thickness. On the other hand, the total biofilm thickness was evaluated by three methods only. This is also to observe that when the HRT (θ) and the SRT (θ_c) are reduced to 50% in case of the method developed by Fouad and Bhargava [13], the effluent substrate concentration (S_w) remains same with respect to the same by Lee [79].

Most of the mathematical models on hybrid bioreactor were firstly illustrated using one or more sets of available data earlier research studies for the purpose of analytical validation. Glimpses of such analytical validation are already presented in Table 2.1 earlier. At the same time, the concerned researchers carried out laboratory-scale study on the respective hybrid bioreactor developed by them. Thus, they compared the experimental observations in respect of relevant parameters with those available from the respective model. A comparison of output parameters, viz. S_w, J, L_e and L_f, derived from various models and concerned experimental observations is presented in Table 2.3. It clearly shows that the experimental results are satisfactorily corroborating the respective model outputs.

References

1. Wanner O, Morgenroth E (2004) Biofilm modeling with AQUASIM. Water Sci Technol 49(1):137–144
2. Eckenfelder Jr WW (2000) Industrial water pollution control 3rd edn. McGraw-Hill Inc, Singapore
3. Sen D, Randall CW (1990) Performance of fixed film media integrated in the activated sludge reactors to enhance nitrogen removal. Water Sci Technol 30:13–24
4. Campbell H, Schnell A (2003) Upgrading activated sludge systems using free floating plastic media [online]. Available from www.hydroxyl.com/papers/waterdown/UpgradeAS.pdf
5. Rutt K, Seda J, Johnson CH (2006) Two years case study of integrated fixed film activated sludge at Broomfield Co, WWTP. In: Proceedings of the water environment federation, Sesson 10, pp 225–239
6. Wang J-L, Wu L-B (2004) Wastewater treatment in a hybrid biological reactor (HBR): nitrification characteristics. Biomed Environ Sci 17:373–379
7. Lewandowski Z (1985) Nitrification process in activated sludge with suspended marble particles. Water Res. 19:535–539
8. Golla PS, Reddy MP, Simms MK, Laken TJ (1994) Three years of full scale captor process operation at Moundsville WWTP, Water Sci Technol 29:175181
9. Wanner J, Kucman K, Grau P (1988) Activated sludge process combined with biofilm cultivation. Water Res 22:207–215
10. Kaindl N, Tillman U, Mobius CH (1999) Enhancement of capacity and efficiency of a biological wastewater treatment plant. Water Sci Technol 40:231–240
11. Rusten B, McCoy M, Proctor R, Siljuden JG (1998) The innovative biofilm moving bed biofilm reactor/solids contact reaction process for secondary treatment of municipal wastewater. Water Environ Res 67:1083
12. Gebara F (1999) Activated sludge biofilm wastewater treatment system. Water Res 33(1):230–238
13. Fouad M, Bhargava R (2005) A simplified model for the steady-state biofilm-activated sludge reactor. J Environ Manage 74:245–253
14. Guang HC, Ju-Chang H, Lo MC (1997) Removal of rate limiting organic substances in a hybrid biological reactor. Water Sci Technol 35:81–89
15. Matsumura M, Yamamoto T, Wang P, Shinabe K, Ya-suda K (1997) Rapid nitrification with immobilized cell using macro-porous cellulose carrier. Water Res 31:1027–1034. https://doi.org/10.1016/S0043-1354(96)00361-2
16. Green M, Ruskol Y, Lahav O, Tarre S (2001) Chalk as the carrier for nitrifying biofilm in a fluidized bed reactor. Water Res 35:284– 290. https://doi.org/10.1016/S0043-1354(00)00239-6
17. Cantet J, Paul E, Clauss F (1996) Upgrading performance of an activated sludge process through addition of talqueous powder. Water Sci Technol 34:75–83. https://doi.org/10.1016/0273-1223(96)00631-2
18. Sakai Y, Tani K, Tkahashi F (1992) Sewage treatment under conditions of balancing microbial growth and cell decay with a high concentration of activated sludge supplemented with ferromagnetic powder. J Ferment Bioeng 74:413–415. https://doi.org/10.1016/0922-338X(92)90045-V
19. Liu JX, Groenestijn JWV, Doddema HJ Wang BZ (1996) Removal of nitrogen and phosphorus using a new biofilm-activated-sludge system. Water Sci Technol 34:315–322. https://doi.org/10.1016/0273-1223(96)00521-5
20. Morper MR (1994) Upgrading of activated sludge system for nitrogen removal by application of the Linpor-CN process. Water Sci Technology 29:167–176
21. Gilligan TP, Morper M (1999) A unique process for upgrading conventional activated sludge systems for nitrogen removal (online). Paper presented at NE WEA, October 1999, available from

22. Harvey PJ, Siviter CL (1999) Use of the suspended carrier process to upgrade wastewater treatment facilities [online]. Available from https://www.purac.net/downloads/articles/1999% 20NZWWA%20Conference%20Paper%Kaldnes_pdf. Accessed on 10 Aug 2003
23. Ødegaard H, Rusten B, Westrum T (1994) A new moving bed biofilm reactor applications and results. Water Sci Technol 29:157–165
24. Rantanen P, Valve M (2003) A hybrid process for biological phosphorus and nitrogen removal-pilot plant experiments [online]. Available from www.vyh.fi/eng/syke/ppd/ws/biorev.htm
25. Dalentoft E, Thulin P (1997) The use of the Kaldnes suspended carrier process in treatment of wastewaters from the forest industry. Water Sci Technol 35:123–130. https://doi.org/10.1016/S0273-1223(96)00923-7
26. Munch EV, Barr K, Watts S, Keller J (2000) Suspended carrier technology allows upgrading high-rate activated sludge plants for nitrogen removal via process intensification. Water Sci Technol 41:5–12
27. Jensen KR 1995 Effects of integrated fixed film activated sludge on nitrogen removal in biological nutrient removal systems. M.S. Thesis, Virginia Tech., Blacksburg, VA
28. Mitta PR (1994) Utilisation of fixed film media in BNR activated sludge. M.S. Thesis, Virginia Tech., Blacksburg, VA
29. Rittmann BE (1982) Comparative performance of biofilm reactor types. Biotechnol Bioeng XXIV:1341–1370
30. Mudliar S, Banerjee S, Vaidya A, Devotta S (2008) Steady state model for evaluation of external and internal mass transfer effects in an immobilized biofilm. Biores Technol 99:3468–3474
31. Raafat M, Diaa O, Monayeri EI, Amer N (2014) Application of hybrid system to upgrade existing wastewater treatment plants: a case study. Int J Sci: Basic Appl Res 14(2):154–168
32. Ross D, Fernandes W, Briggs T, Kim N, Booh G, Neely D, Welp J (2004) Integrated fixed film activated sludge (IFAS) at the lake view WWTP the real implementation issues. In: Proceedings of the 77 th WEFTEC-national conference of the water environment federation, New Orlians, L.A
33. Rostron WM, Stuckey DC, Young AA (2001) Nitrification of high strength ammonia waste waters: comparative study of immobilization media. Water Res 35:1169–1178. https://doi.org/10.1016/S0043-1354(00)00365-1
34. Chuang SH, Ouyang CF, Yuang HC, You SJ (1997) Effects of SRT and DO on nutrient removal in a combined AS-biofilm process. Water Sci Technol 36:19–27
35. Hait S, Mazumder D (2008) Scope of improvement of treatment capacity of activated sludge process by hybrid modification. J Environ Eng Sci 7:147–158 (NRC Canada)
36. Ye J, McDowell CS, Koch K, Kulick FM, Rothermal BC (2009) Pilot testing of structured sheet media IFAS for wastewater biological nutrient removal(BNR). Proc Water Environ Fed 12:4427–4442
37. Gheewala SH Annachhatre AP (2003) Efficacy of biofilms under toxic conditions. J Env Engg ASCE 576–579
38. Mazumder D (2010) Simultaneous COD and ammonium nitrogen removal from a high strength wastewater in a shaft-type aerobic hybrid bioreactor. Int J Environ Sci Dev 1(4):327332
39. Li C, Ji M, Li X, Wang M (2011) Municipal wastewater treatment in a new type bio-carrier reactor. Procedia Environ Sci 10:962–967
40. Hooshyari B, Azimi A, Mehrdadi N (2009) Kinetic analysis of enhanced biological phosphorus removal in a hybrid integrated fixed film activated sludge process. Int J Environ Sci Technol 6(1):149–158
41. Azimi AA, Hooshyari B, Mehrdadi N, Bidhendi GN (2007) Enhanced COD and nutrient removal efficiency in a hybrid integrated fixed film activated sludge process. Iran J Sci Technol 31:523–533
42. El-jafry MH, Ibrahim WA, El-adawy MS (2013) Enhanced COD and nutrient removal efficiency in integrated fixed film activated sludge (IFAS) process. In: International conference on civil and architecture engineering, Kuala Lampur, pp 43–46
43. Thalla AK et al (2012) Experimental evaluation of activated sludge—biofilm reactor: influence of composite media. ASCE, HZ 2153–5515:134–141

44. Lobos-Moysa E (2012) Application of hybrid biological techniques to the treatment of municipal wastewater containing oils and fats. Desalin Water Treat (impact factor: 0.61). https://doi.org/10.1080/19443994.2012.677451
45. Arumugam SM, Ponnusami V (2013) Aerobic treatment of dairy wastewater by hybrid bioreactor. Int J Chem Res 5(6):3064–3069
46. Bohdziewicz J, Sroka E (2006) Application of hybrid systems to the treatment of meat industry wastewater. Desalination 198:33–40
47. Farrokhi M, Ashrafi D, Roohbakhsh E, Yoonesi A (2014) Hospital wastewater treatment by integrated fixed film activated sludge, using rice husk as fixed media. Adv Life Sci 4(3):178–183
48. Moysa EL, Dudziak M, Zoń Z (2009) Biodegradation of rapeseed oil by activated sludge method in the hybrid system. Desalination 241:43–48
49. Trapani DD, Christensso M, Odegaard H (2011) Hybrid/activated sludge /biofilm process for the treatment of municipal wastewater. Water Sci Technol 63(6):1121–1129. https://doi.org/10.2166/wst.2011.350
50. Oles J, Wilderer PA (1991) Computer aided design of sequencing batch reactors based on the IAW PRC activated sludge model. Water Sci Technol 23(4–6): 1087–1095
51. Daigger GT, Nalosco D (1995) Evaluation and design of full-scale wastewater treatment plant using biological process models. Water Sci Technol 31(2):245–255
52. Mehrdadi N, Azimi A, Bidhendi GRN, Hooshyari B (2006) Determination of design criteria of an H-IFAS reactor in comparison with an extended aeration activated sludge process. Iran J Environ Health Sci Eng 3(1)
53. Finnson A (1993) Simulation of a strategy to start up nitrification at Bromma sewage plant using a model based on IAW PRC model No.1. Water Sci Technol 28(11–12):185–195
54. Novotny V, Jones HV, Feng X, Capodaglio AG (1990) Time series analysis models of activated sludge plants. Water Sci Technol 23:1107–1116
55. Capodaglio AG, Jones HV, Novotny V, Feng X (1991) Sludge bulking analysis and forecasting: application of system identification and artificial neural computing technologies. Water Res 25(10):1217–1224
56. Cote M, Gransjean BPA, Lessard P, Thibault J (1995) Dynamic modeling of the activated sludge process: improving prediction using neural networks. Water Res 29:995–1004
57. Lukasse LJS, Keesman KJ, Klapwijk A, van Straten G (1988) Optimal control of N-removal in ASPs. Water Sci Technol 38(3):255–262
58. Lesouef A, Payraudeau M, Rogalla M, Kleiber B (1992) Optimizating nitrogen removal reactor configurations by on-site calibration of the IAW PRC activated sludge model. Water Sci Technol 25(6):105–123
59. Shaw AR, Johnson TL, Johnson C (2003) Intricacies of modeling the emerging integrated fixed film activated sludge (IFAS) process. In: Proceedings of the water environment federation, WEFTEC 2003: Session 61 through Session 70, pp 95107(13)
60. Sriwiriyarat T, Randall CW (2005) Evaluation of integrated fixed film activated sludge wastewater treatment processes at high mean cells residence time and low temperatures. J Environ Eng ASCE 131(11):1550–1556
61. Harris NP, Hansford GS (1976) A study of substrate removal in a microbial film reactor. Water Res 10: 935–943
62. Henze M, Grady Jr CPL, Gujer W, Marais GVR, Matsuo T (1987) Activated sludge model No. 1. Scientific and Technical report No. 1. International Association on Water Pollution Research and Control, London
63. Barker PS, Dold PL (1997) General model for biological nutrient removal activated sludge systems. Water Environ Res 69(5):969–984
64. Henze M, Gujer W, Mino T, van Loosdrecht M, Marais, GVR, Wentzel MC, Matsuo T (1999) Activated sludge model No. 2D, ASM 2D. Water Sci Technol 39(1):165–182
65. Tsuno H, Hidakaa T, Nishimura F (2001) A simple biofilm model of bacterial competition for attached surface. Water Res 36:996–1006
66. Boltz JP, Morgenroth E, Sen D (2010) Mathematical modeling of biofilms and biofilm reactors for engineering design. Water Sci Technol 62:1821–1836

67. Williamson K, McCarty PL (1976) A model of substrate utilization by bacterial films 48(1):9–23
68. Rittmann BE, McCarty Perry L (1980) Model of steady state biofilm kinetics. Biotechnol Bioeng XXII:2343–2357
69. Huang JS, Jih CG (1997) Deep biofilm kinetics of substrate utilization in anaerobic filters. Water Res 31(9):2309–2317
70. Kim BR, Suidan MT (1988) Approximate algebraic solution for a biofilm model with the monod kinetic expression. Water Res 23(12):1491–1498
71. Heath MS, Wirtel SA, Rittmann BE (1990) Simplified design of biofilm processes using normalized loading curves. Water Pollution Control Federation 62:185–191
72. Saez PB, Rittmann BE (1991) Accurate pseudo analytical solutions for steady state biofilm. Biotech Bioengg 39:790–793
73. Pritchett LA, Dockery JD (2001) Steady state solutions of a one dimensional biofilm model. Math Comput Modell 33:255–263
74. Hsien TY, Lin YH (2005) Biodegradation of phenolic wastewater in a fixed biofilm reactor. Biochem Eng J 27:95–103
75. Perez J, Picioreanu C, Loosdrecht MV (2005) Modeling biofilm and flock diffusion processes based on analytical solution of reaction-diffusion equations. Water Res 39:1311–1323
76. Wik T, Goransson E, Breitholtz C (2006) Low model order approximations of continuously stirred biofilm reactors with Monod kinetics. Biochem Eng J 30:16–25
77. Eswari S, Modi A, Kapoor DP, Venkateswarlu C (2012) Modelling of fixed bed biofilm reactor treating pharmaceutical industry wastewater. In: Conference paper in chemical engineering sciences, IICT, Hyderabad
78. Sarkar S, Mazumder D (2015) Process design and application of aerobic hybrid bioreactor in the treatment of municipal wastewater. Int J Chem Nucl Mater ialsand Metall Eng World Acad Sci Eng Technol 9(3):466–470
79. Lee C-Y (1992) Model for biological reactors having suspended and attached growths. J Environ Eng 118(6):982–987
80. Sriwiriyarat T (1999) Computer programme development for the design of IFAS wastewater treatment processes. M.S.Thesis, Virginia Tech, Blacksburgs, VA 24061, pp 1540–1549
81. Fadel M, Ibrahim A, Ayoub G (2002) Aerobic hybrid growth process modeling: a parametric sensitivity analysis. In: Proceedings of international symposium on environmental pollution control and waste management, pp 333–348
82. Pai TY, Chuang SH, Tsai YP, Yuyang CF (2004) Modelling a combined anaerobc/anoxic oxide and rotating biological contactors process under dissolved oxygen variation by using an activated sludge-biofilm hybrid model. J Env Eng 130(12):1433–1441
83. Sriwiriyarat T et al (2005) Computer program development for the design of integrated fixed film activated sludge wastewater treatment processes. J Environ Eng 131(11):1540–1549
84. Boltz JP, Johnson BR, Daigger GT, Sandino J (2009) Modeling integrated fixed-film activated sludge and moving- bed biofilm reactor systems i: mathematical treatment and model development. Water Environ Res 81(6):555–575
85. Metcalf L, Eddy HP (2013) Wastewater engineering. A text book, Reprint 21st
86. Majdabadi MM (2011) New modification for wastewater treatment plant. Int Conf Environ Agricul Eng IPCBEE 15:130–133
87. Mardani S et al (2011) Determination of biokinetic coefficients for activated sludge processes on municipal wastewater. Iran J Environ Health Sci Eng 8(1):25–34
88. Daniel D, Christensso M, Odegaard H (2011) Hybrid/activated sludge/biofilm process for the treatment of municipal waste water. Water Sci Technol 63(6):1121-1129. doi: https://doi.org/10.2166/wst.2011.350
89. Hinson RK, Kocher WM (1996) Model for effective diffusivities in aerobic biofilms. ASCE J Environ Eng ASCE 122(11):1023–1030

90. Lee CY, Lan YW (1999) Competitiveness evaluation for individual growth in hybrid processes. J Environ Eng ASCE 125(2):146–152
91. Rittmann BE, McCarty PL (1980) Evaluation of steady-state-biofilm kinetics. Biotech Bioeng XXII:2359–2373

Chapter 3
Materials and Methods

3.1 Approach for Modeling of Fixed-Bed Hybrid Bioreactor

A simplified model for fixed-bed hybrid bioreactor is developed for easily determining the output parameters like exiting substrate concentration in bulk liquid (S_w), average substrate flux in the biofilm (J_{avg}) and effective as well as total biofilm thickness (L_e and L_f). The conceptual diagram of a typical fixed-bed hybrid bioreactor comprising of activated sludge–biofilm is shown in Fig. 3.1.

In the above completely mixed hybrid bioreactor system, a portion of substrate is uniformly exposed to the suspended biomass. The remaining fraction of substrate flows through the biofilm–liquid interface and then through the biofilm. Thus, the substrate mass balance for the hybrid bioreactor is obtained by the combination of substrate mass balance equations for both suspended and attached-growth biomass with their individual fraction of substrate uptake in their respective growth. For biomass balance of the hybrid bioreactor, the portion of shear loss of biofilm from the attached growth is also taken into consideration. Before discussion for approaches in mathematical modeling of hybrid bioreactor, an approach of modeling each for suspended and attached growth is highlighted as follows.

3.1.1 *Modeling of Suspended-Growth System*

The approach of mathematical model for suspended-growth reactor is based on substrate mass balance and biomass balance equations using Monod kinetics in substrate utilization by the suspended biomass. The following equation is the substrate mass balance in the suspended growth,

$$(S_0 - S_w) - \frac{pkXS_w\theta}{K_s + S_w} = 0$$

S. Sarkar and D. Mazumder, *Fixed Bed Hybrid Bioreactor*, Green Energy and Technology,
https://doi.org/10.1007/978-981-33-4546-1_3

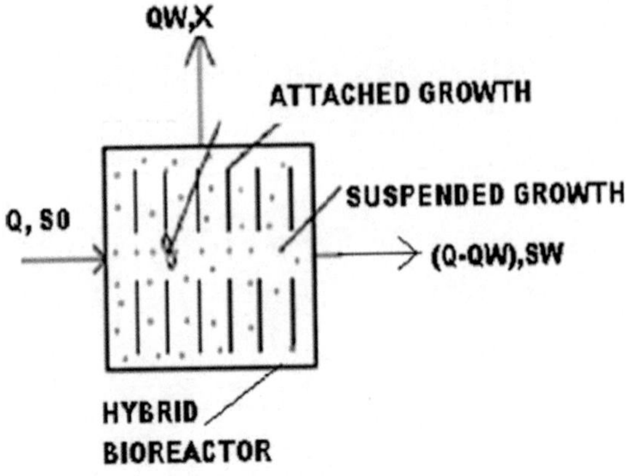

Fig. 3.1 Schematic diagram of a hybrid bioreactor

where S_0 = substrate concentrations in the bulk liquid (mg/cm^3), S_w = effluent substrate concentration (mg/cm^3), X = biomass concentration in suspended growth (mg/cm^3), k = maximum specific rate of substrate use (per day), θ = empty-bed hydraulic detention time (day), K_s = half velocity coefficient (mg/cm^3) and p = porosity of hybrid reactor.

The mass balance equation for suspended-growth biomass is as follows:

$$\frac{1}{\theta_c} - \frac{YkS_w}{K_s + S_w} + b_d = 0$$

where θ_c = mean cell residence time or solid retention time (day^{-1}).

Y = bacteria yield coefficient,

b_d = biomass decay coefficient, day^{-1}.

3.1.2 *Modeling of Biofilm Growth System*

Modeling of biofilm growth can be approached in two ways depending on the condition of the biofilm reactor, i.e., whether it is a fixed biofilm reactor or it is a completely mixed biofilm reactor. In case of fixed biofilm reactor, mathematical model was developed by formulating relationship between the substrate concentrations at initial, both the entry and exit, at biofilm–liquid interface and at biofilm attached surface along with average substrate flux in the biofilm, substrate flux at biofilm–liquid interface and effective biofilm thickness. However, in case of completely mixed biofilm reactor, mathematical model was developed formulating correlations between entering and exiting substrate concentrations and that at biofilm attachment surface along with average substrate flux in the biofilm and effective biofilm thickness. The profile

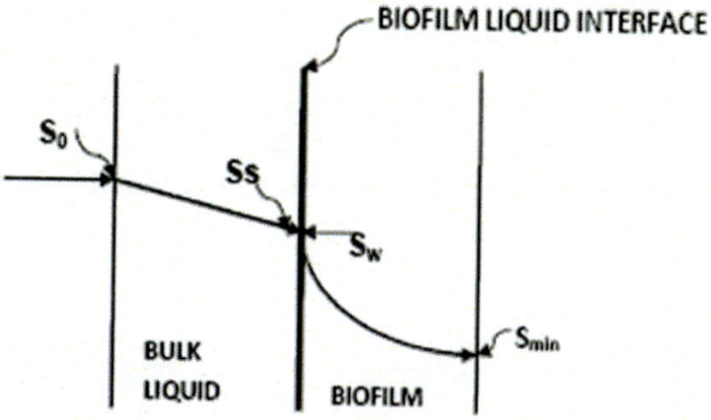

Fig. 3.2 Profile of substrate concentration in a fixed biofilm reactor (FBR)

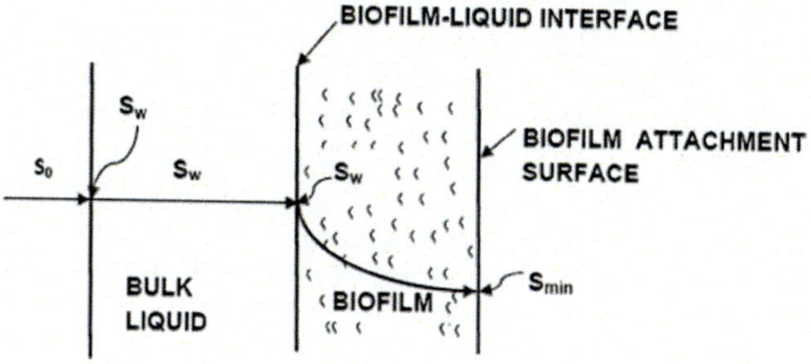

Fig. 3.3 Profile of substrate concentration in a completely mixed biofilm reactor (CMBR)

of substrate concentration in a fixed biofilm reactor and completely mixed biofilm reactor is shown in Figs. 3.2 and 3.3, respectively.

From Fig. 3.2, it is evident that entering substrate concentration S_0 in the bulk liquid becomes entering substrate concentration S_s and exiting substrate concentration S_w at the biofilm–liquid interface due to diffusion in the stagnant liquid layer attached to biofilm, and it further decreases to S_{min}, i.e., the minimum substrate concentration at biofilm attachment surface. The substrate flux which is varying over the biofilm thickness can be averaged to account for the substrate mass balance. This average flux must be same as that derived from the entering substrate concentration S_s and the exiting substrate concentration S_w in the biofilm–liquid interface.

From Fig. 3.3, it is evident that entering substrate concentration S_0 becomes exiting substrate concentration S_w in the bulk liquid due to mechanical mixing in the reactor. This concentration S_w is the homogeneous bulk liquid concentration up to the liquid–biofilm interface and which further decreases to S_{min}, i.e., the minimum substrate concentration at biofilm attachment surface. There is a variation in the substrate flux over the biofilm thickness that can be averaged to account for the substrate mass balance. This average flux must be same as that derived from the entering substrate concentration S_0 and the exiting substrate concentration S_w in the biofilm reactor. The mathematical model for the biofilm growth essentially considered the substrate mass transport external to the biofilm and into the biofilm as per Fick's law and steady-state substrate as well as biomass balance for biofilm. Monod's growth kinetics has been adopted in substrate utilization incorporating relevant boundary conditions. The numerical solution of model equations was accomplished for calculating average flux and exiting substrate concentration and thereafter applied Runge–Kutta method for determining effective biofilm thickness.

3.1.3 Modeling of Hybrid Growth System

Modeling of hybrid growth can be approached considering mass balance of both carbonaceous substrate and biomass under suspended and attached growth simultaneously along with substrate mass transport into the biofilm. Monod kinetics is followed for the utilization of carbonaceous substrate assuming no inhibition. In this modeling approach, a fraction of substrate used by suspended biomass at steady-state condition is considered as r and that used by attached biomass as $(1 - r)$. Kinetic equation for the hybrid bioreactor is derived from the combination of the two equations, each of substrate mass balance of both suspended-growth biomass and attached-growth biomass with the fraction of substrate utilized as r and $(1 - r)$, respectively.

Profile of substrate concentration in a fixed-bed hybrid bioreactor is shown in Fig. 3.4.

Hybrid bioreactor is considered as completely mixed, and thus, it is evident from Fig. 3.4 that entering substrate concentration S_0 becomes exiting substrate concentration S_w in the bulk liquid. This concentration S_w is the homogeneous bulk liquid concentration up to the liquid–biofilm interface as in case of completely mixed biofilm reactor. It then further decreases to S_{min}, i.e., the minimum substrate concentration at biofilm attachment. Since the hybrid bioreactor is a combination process of both the suspended and attached-growth biomass, average substrate flux is also considered inside the biofilm in this hybrid bioreactor model; likewise, biofilm growth modeling is considered earlier. Average substrate flux can be determined from the solution of classical mass balance equation of substrate in biofilm. In the combined

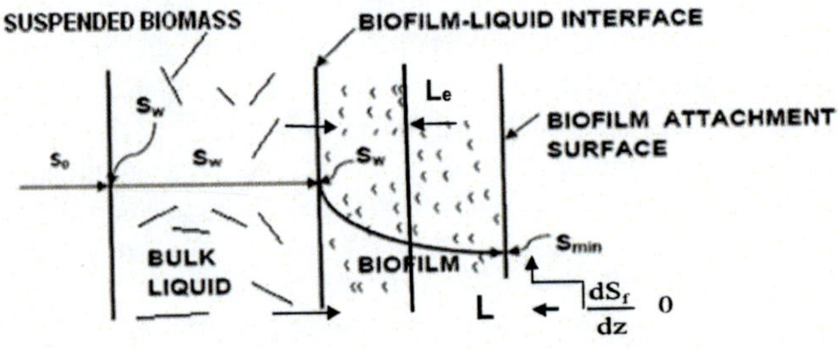

Fig. 3.4 Profile of substrate concentration in a hybrid bioreactor

substrate mass balance equation developed from both suspended and attached growth, the suspended biomass concentration can be substituted by the expression derived from the steady-state mass balance of active micro-organisms in a biofilm as well as under suspended-growth state. While biomass balance equation is exercised for hybrid bioreactor, the shear loss of biofilm from the attached growth has been added to the suspended-growth biomass. Now, the effective biofilm thickness can be determined by applying Runge–Kutta method as a solution of classical mass balance equation of substrate in the biofilm.

3.2 Approach for the Development of Analytical Procedure for Determining the Kinetic Coefficients of Hybrid Bioreactor

In order to determine the kinetic coefficients of hybrid bioreactor, analytical solution has been done in the following steps:

Step 1

$$\mu = \frac{(X_{2a} + X_{2s}) - (X_{1a} + X_{1s})}{(X_{1a} + X_{1s})\theta}$$

μ = specific growth rate of biomass for hybrid bioreactor (h^{-1}).

X_{1a} = initial biomass density for attached phase (mg/cc).

X_{1s} = initial biomass density for suspended phase (mg/cc).

X_{2a} = final biomass density for attached phase at the end of batch period (mg/cc).

X_{2s} = final biomass density for suspended phase at the end of batch period (mg/cc).

θ = batch period under consideration (h).

Biomass density can be determined by washing the biofilm attached surface with 25 mL 0.1 (N) NaOH after heating the solution at 80 °C [1, 2] and subsequently by measuring the protein concentration with Lowry's method.

Step 2

The 'μ' values can be plotted in Y-axis with respect to 'S_w' values in X-axis, which should have an ascending asymptotic nature as shown in Fig. 4c of Annexure I-b. The half-velocity constant K can be obtained corresponding to half of the maximum specific growth rate, i.e., μm, as indicated in the same figure. The specific growth curve of biomass as shown in Fig. 4c of Annexure I-b essentially represents the Monod kinetics, which is valid for non-inhibitory environment. In case of inhibitory environment, the specific growth curve will reach to a maximum value (i.e., μm), and then, it will fall downward.

Step 3

From Eq. 4.15 in Chap. 4,

$$\frac{pkXS_w\theta}{(S_0-S_w)(K+S_w)} + \frac{aJ\theta}{(S_0-S_w)} = 1$$

$$\frac{\{(X\theta S_w)/(S_0-S_w)(K+S_w)\}}{1/pk} + \frac{\theta/(S_0-S_w)}{1/aJ} = 1$$

Now, $\left\{\frac{X\theta S_w}{(S_0-S_w)(K+S_w)}\right\}$ values can be plotted with respect to $\left\{\frac{\theta}{(S_0-S_w)}\right\}$ and a best-fit line can be drawn as shown in Fig. 4d, Annexure I-b.

Let $1/(pk)$ = intercept in Y-axis (m) and $1/aJ$ = intercept in X-axis (n), Hence, $k = 1/pm$ and $J = 1/an$.

Again, $k = \mu m/Y \geq Y = \mu m/k$.

The maximum specific substrate utilization rate (k) can be determined from the Y-intercept of Fig. 4d of Annexure I-b, provided the porosity of the hybrid bioreactor in terms of attached phase is known. On the other hand, the 'J' value evaluated from the X-intercept of Fig. 4d of Annexure I-b represents the mean substrate flux through the attached biofilm in the hybrid bioreactor.

Step 4

Equation 4.14 in Chap. 4 yields the biomass balance for the suspended-growth system as,

$$=> \frac{YaJ_{avg}}{pX}\left(\frac{b_s}{b_s+b_d}\right) = \frac{1}{\theta_c} - \frac{YkS_w}{K+S_w} + b_d$$

$$=> \frac{YaJ_{avg}}{pX} = \left(\frac{1}{\theta_c} - \frac{YkS_w}{K+S_w}\right)\left(\frac{b_s+b_d}{b_s}\right) + b_d\left(\frac{b_s+b_d}{b_s}\right)$$

Now, $\frac{YaJ}{pX}$ values can be plotted with respect to $\left(\frac{1}{\theta_c} - \frac{YkS_w}{K+S_w}\right)$ values and a best-fit line can be drawn as shown in Fig. 3.5.

This mean value of 'J' is used to calculate $\frac{YaJ_{avg}}{pX}$ as shown in Fig. 3.5.

Let $c = b_d\left(\frac{b_s+b_d}{b_s}\right) = b_d$ m, which follows,

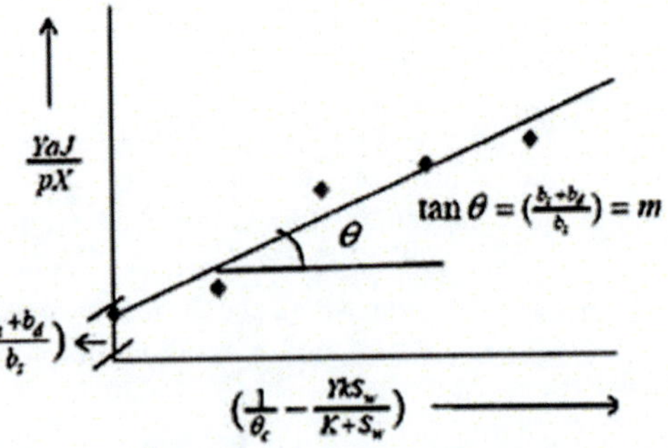

Fig. 3.5 Biomass balance for suspended growth for the determination of b_s, b_t, b_d

$b_d = c/m$ and $b_s = b_d/(m - 1) = c/\{m(m - 1)\}$

$$b_t = \frac{c}{(m - 1)}$$

Although 'θ_c' represents the solid retention time for the suspended biomass, it becomes $p * \theta$ in a semi-batch reactor provided with no recirculation. In this case, also the slope and intercept of the 'best-fit line' considerably influence the values of b_s and b_d.

3.3 Practice of Aeration in the Reactor

The aeration facility has been provided in the reactor with air compressor along with controlled airflow through airflow meter with an average airflow of 20–30 l/m. The airflow should be uniformly distributed through air nozzles to different slots of attached media, i.e., through nine (9) number slots in octagonal attachment media used in our experiment. Mechanical mixing with impeller along with aeration facility maintains proper agitation in order to make availability of suspended MLSS in the aeration tank for the biological activity. The DO concentration maintained in the reactor is between 0.5 and 1 mg/l. Diffused aeration is employed with submerged course diffusers.

3.4 Sampling Procedure and Preparation of Sample

3.4.1 Sampling Procedure

The hybrid bioreactor was run by both synthetic municipal wastewater and raw municipal wastewater under continuous mode. In case of synthetic municipal wastewater, test was run with three variable initial COD concentrations, viz. 150, 200 and 250 mg/l and maintaining a HRT of 4, 6 and 8 h. The initial biomass concentration (both suspended and attached) was also varied as per initial COD concentration. The reactor was allowed to run by means of peristaltic pump until quasi-steady-state condition is reached. During the continuous study, the effluent sample was taken from the overflow of the secondary clarifier at an interval of three (3) h for measuring the final COD concentration of its soluble sample. The reactor content was taken for determining suspended biomass concentration at both initial and final conditions. A portion of attached media (a detachable perspex sheet fin) was washed out with 1 N NaOH solution to measure the attached biomass.

3.4.1.1 Determination of Protein Concentration [3, 4]

Apparatus: UV–Vis spectrophotometer, conical flask, glass beaker and pipette.

Reagents

(a) 5% Na_2CO_3 solution (5 g of Na_2CO_3 per 100 mL of distilled water).
(b) $CuSO_4$ solution (0.5 mg of $CuSO_4$, $5H_20$ is dissolved in 100 mL of 1% Na-K-tartrate solution).
(c) Alkali copper reagent (50 mL of reagent a + 2 mL of reagent b). The reagent should be prepared freshly just before use.
(d) Folin reagent solution (Folin reagent: distilled water = 1:1).
(e) 1 (N) NaOH solution (4 g NaOH is dissolved in 100 mL distilled water).
(f) Standard bovine serum albumin (BSA) solution of concentration 200 mf/l (20 mg of BSA, Sigma USA brand is dissolved in 100 mL distilled water).

Procedure
The cell suspension of 0.5 mL was mixed with 0.5 mL 1 (N) NaOH solution and kept in boiling water bath for 5 min. The content was then cooled in cold water. A 5 mL of freshly prepared reagent C was added to the content and allowed to stand for 10 min. Now, 0.5 mL of Folin–Ciocalteu's reagent solution was added and the whole mixture was allowed to stand for 30 min for color development. The blank was prepared taking 0.5 mL of distilled water instead of bacterial suspension and treating it in the same way. The absorbance value was measured at 750 nm wavelength using a visible spectrophotometer against the distilled water blank. BSA standards ranged between 0 and 200 μg/mL were processed in the same manner as samples for developing the calibration curve.

3.4.1.2 Biological Sample for Observation in Scanning Electron Microscope Was Prepared in Following Ways

Fixation: Biofilm is attached to the perspex sheet (1 cm × 1 cm, 3 mm thickness) in the treatment process.

- Dehydration: Successive immersion of the biofilm surface in ethanol solution of varying strengths like 10, 25, 50, 75, 96 and 99% for 30 min in each case.
- Drying: At 37 degree centigrade for 1 h.
- Coating: The prepared samples are to be coated with a thin gold film (<10 nm).
- Acceleration voltage maintained: <1 kV.
- Working distance: <1 mm.

3.4.2 *Methods of Sample Analysis*

All the relevant parameters except MLSS were measured as per Standard Methods [5]. In case of MLSS, 50 mL of mixed liquor sample was taken out from the reactor and the same was filtered. The filtrate retained in the filter paper was allowed to dry in a hot air oven at 105° C until it attains a constant weight. Then, the filter paper was put into a desiccator, and its final weight was taken. Thus, it is possible to calculate the MLSS by subtracting the initial weight of the filter paper from its final weight. To determine the COD of the soluble sample of the effluent, the effluent was passed through the filter paper and the filtrate sample was taken for determining its COD by closed reflux method [5].

Biomass density in the attached media was determined by washing the biofilm surface with 25 mL 0.1 (N) NaOH and then heating the solution at 80 °C [1, 2]. Subsequently, by measuring the protein concentration in the washed out content as per Lowry's method [4], the amount of attached biomass was estimated. Biofilm thickness was measured in scanning electron microscope.

3.5 Water Used in Experimental Study

Double distilled water was used in preparation of all chemical reagents needed for carrying out the laboratory tests. Single distilled water is used for washing all necessary glass wares, test tubes and vials after acid washing followed by the washing with tap water.

3.6 Chemicals and Glass Wares Used in Experimental Study

All chemicals used for preparation of reagents are of "Analytical Grade" and manufactured by E-Merck, Germany.

Conical flasks, measuring cylinders, beakers, funnels and other glass wares including vials, test tubes, etc., were made of Borosilicate.

3.7 Instruments Used in Experimental Study

The instruments used for the determination of various analytical parameters are listed in Table 3.1.

Table 3.1 List of instruments used in the laboratory

Name of equipment	Make and model
Double beam UV-VIS Spectrophotometer	Systronics
COD digestion apparatus (Thermoreactor)	E-Merck
BOD Incubator	Instrumentation India
DO Meter	Orion
ION Meter	Orion Star A214
Scanning Electron Microscope	Hitachi S 3400 N
Hot Air Oven	Instrumentation India,
Desiccators	Tarsons
Weighing M/c	Afcoset
Muffle Furnace	SAI
BOD Shaker Incubator	Modern Instrument

3.8 Origin and Enrichment of Biomass

Biomass used in the experimental reactor was firstly taken from an activated sludge process unit. Thereafter, it was enriched under acclimation phase using real municipal wastewater. Biomass sample was taken from both suspended portion and attached portion after their growth at steady state in the reactor for their isolation, identification and DNA study.

3.9 Methodology of Bacterial Identification

3.9.1 Isolation of Pure Cultures

3.9.1.1 Sterilization

Distilled water, 10 test tubes with 9 mL of distilled water each, 10 Petri dishes, nutrient agar, 250-mL conical flask with 99 mL of distilled water are sterilized in auto-clave under 15 lb/sq. inch pressure.

3.9.1.2 Serial Dilution

At first, 1 g of active sludge sample taken from the reactor was mixed in 99 mL of sterile distilled water in the sterilized 250-mL conical flask. After cotton plugging, the flask was shaken for 1 h. Then, sterilized test tubes containing 9 mL of sterile distilled water each and marked as 10^{-1}–10^{-9} were taken. 1 mL of sample solution was taken out using pipette from the conical flask and was mixed with 10-1 test tube. Again 1 mL diluted sample was taken from 10-1 marked test tube and was mixed with 10-2 marked test tube. Like this, serial dilution process for 10^{-1} to 10^{-9} dilution in nine (9) test tubes was performed.

3.9.1.3 Preparation of Agar Plate and Spreading

Composition of Agar Medium

- Beef extract—1.5 gm/500 mL
- Peptone—2.5 gm/500 mL
- Sodium chloride—2.5 gm/500 mL
- Distilled water—500 mL
- Agar—8 gm/500 mL
- pH—7.0–7.2.

All compositions were mixed well in a 1000-mL beaker and were sterilized in an auto-clave under 15 lb/sq. inch pressure at saturated steam condition and 121 °C temperature for 20–25 min. 10 mL of sterile agar medium was poured into each 9 number of sterile Petri dishes. After solidifying, the plates were marked as 10^{-1}–10^{-9}, respectively. Then, 100 micro-liters of each sample dilution from test tubes was poured and spread over the respective dilution marked Petri dish. Thus, the dilution sample was poured and spread over 10^{-1} to 10^{-9} marked Petri dishes. Then, the Petri dishes were incubated at 37 °C for 24 h.

3.9.1.4 Observations

Colonies were developed upon the Petri dishes within the incubation period. The colonies were counted and marked in corresponding Petri dishes.

3.9.1.5 Gram Staining of Bacteria

Gram Staining Theory

One of the most important and widely used differential staining techniques in microbiology is gram staining. This technique was introduced by Christian Gram in 1884. In this process, the fixed bacterial smear is subjected to the following staining reagents in the order listed: crystal violet, iodine solution, alcohol (decolorizing agent) and staining. Gram-positive bacteria retain the crystal violet and hence appear deep violet in color, and Gram-negative bacteria, which loss the crystal violet, are counterstained by the safranin and hence appear red in color.

Gram Staining Procedure of Bacteria

1. One clean slide was gently heated in the flame to remove any grease present.
2. A loopful of distilled water was placed in the center of the slide.
3. A minute amount of bacterial growth with the help of a wire loop was removed.
4. The bacterial growth was emulsified in a loopful of distilled water on the slide. The smear was prepared by spreading the suspension in a thin film on the slide.
5. The smear was fixed quickly passing the slide through the flame.
6. The film was covered with sufficient crystal violet stain and allowed to remain for 1 min.
7. The excess stain was passed off and washed the slide with tap water.
8. The film was covered with Gram's iodine solution and was allowed to stand for 1 min.
9. The slide was washed on tap water and dried.
10. 95% of ethyl alcohol was passed dropwise till the violet color flowed away.
11. The slide was covered with Safranin for 30 s.
12. The slide was washed in tap water and blotted dry.
13. Then, the slide was examined under microscope.

3.9.1.6 Bacterial Characterization Using 16S rDNA Protocol

Genomic DNA was extracted from the bacterial culture both suspended and attached by conventional method, and16S rDNA amplification was carried out with 27F and 1492R universal primer sequences. The capillary sequencing was done by ABi 3730XL Genetic Analyzer machine as per manufacture's information.

Amplification Strategy:

Sample:	1 μL	(100 times dilution of genomic DNA)
Primer (10 pmol/μL):	0.5/0.5 μL	(27F/1492R)
Taq Master mix (G9 Taq, GCC Biotech):		25 μL
Distilled water:		23 μL
Total volume:		50 μL

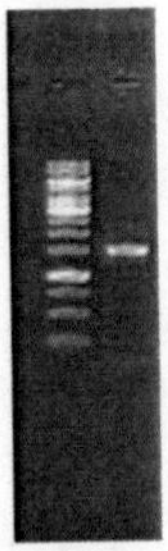

Agarose Gel Pic (1% agarose): Lane1 to Lane2: 1 kb ladder (1 μL), A1 (1 μL).

PCR Cycle:

95.0 °C—10 min

95.0 °C—30 s

54.0 °C—1 min 30 cycles

72.0 °C—1:30 min

72.0 °C—10 min

4.0 °C— □

The results of phylogenetic tree are shown in Figs. 3.6 and 3.7.

3.10 Analytical Methods

3.10.1 Chemical Oxygen Demand (COD)

COD of the samples was measured following the principle of "Closed Reflux Method." At first, 2.5 mL of sample or its suitable aliquot was taken to a COD culture tube and added with 1.5 mL of 0.1 N $K_2Cr_2O_7$ and 3.5 mL of Ag_2SO_4 containing H_2SO_4 one after another. This culture tube solution was denoted as sample solution. A similar tube was prepared taking 2.5 mL of distilled water in place of sample, which is named as blank solution. Now, these two COD culture tubes were allowed to digestion in a refluxing apparatus (COD Thermoreactor) at about $(148 \pm 2)^{\circ}$ C for 2 h. After the digestion was over, the tubes were cooled and their contents are titrated with 0.01 N ferrous ammonium sulfate (FAS) using ferroin indicator until a reddish brown color appears. To determine the strength of FAS solution, a standardization was also performed using a 'blank' tube as prepared earlier. In this case, the culture tube was directly titrated with FAS without any digestion in the refluxing apparatus.

$$\text{COD} = \{(A - B) \times M \times 8000/\{\text{volume of sample}\}$$

A = Volume of FAS for blank solution.

B = Volume of FAS for sample solution.

M = Calculated morality of FAS.

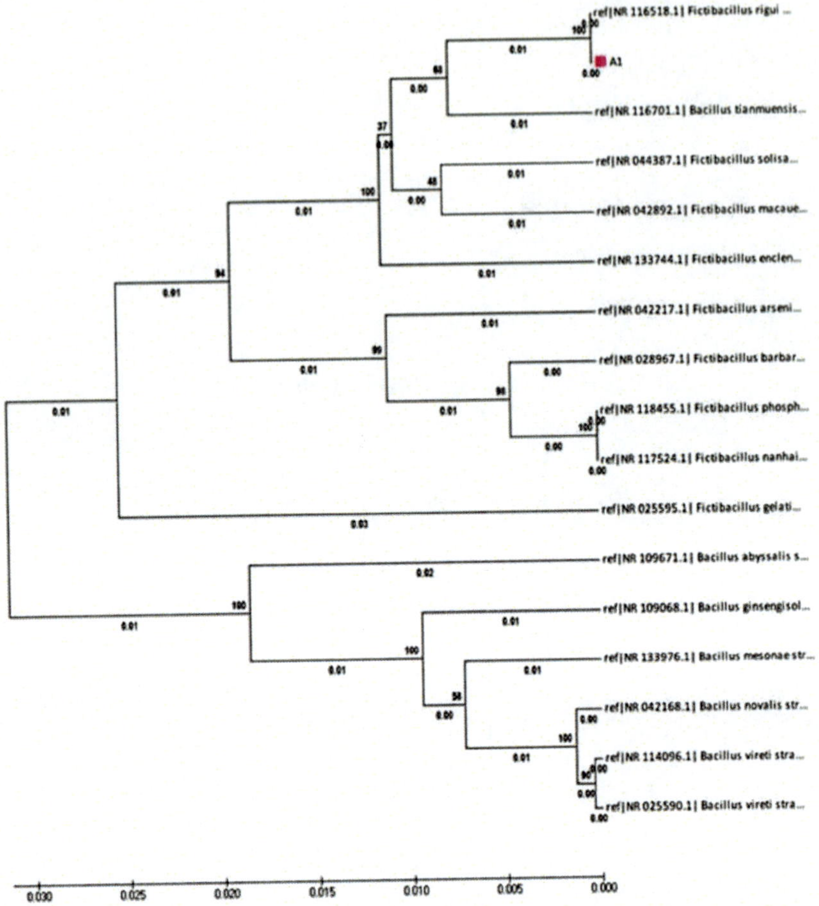

Fig. 3.6 Unrooted phylogenetic tree based on 16S rDNA sequences obtained by neighbor-joining (NJ) method showing the position of strain *Fictibacillus rigui* among its phylogenetic neighbors, NCBI accession numbers are provided in parenthesis

3.10.2 *Mixed Liquor Suspended Solids (MLSS)*

At first, the weight of commercial filter paper is taken by the weighing machine. 50 mL of uniformly mixed suspended biomass solution was taken by measuring cylinder and filtered through commercial filter paper placed in a funnel. The contents of biological reactor (biomass) were mechanically stirred gently for 1–2 min, and the sample was taken immediately from the middle portion of the stirred biomass to maintain uniformity in the mixture. The time taken for the above procedure was preferred as less as possible so that no settlement of suspended particles occurs, in other words prevention of non-uniformity of sample solution. The filter paper along with the residue on it was kept in an oven at 105 °C for about 2 h until the moisture

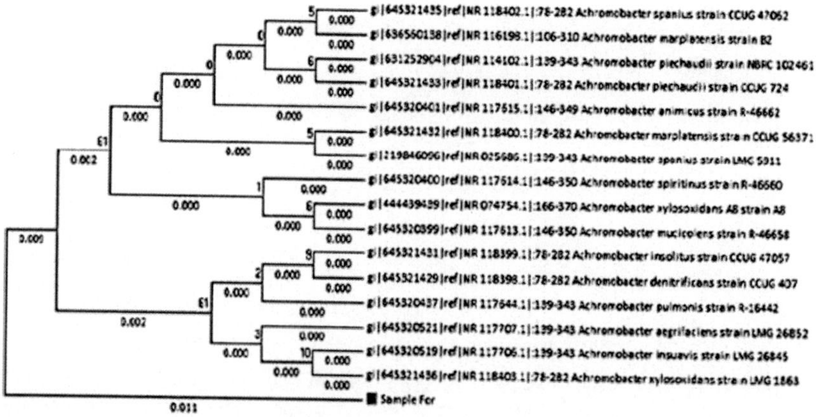

Fig. 3.7 Unrooted phylogenetic tree based on 16S rDNA sequences obtained by neighbor-joining (NJ) method showing the position of strain *Achromobacter xylosoxidans* A8 strain

got removed. The oven-dried filter paper along with residue was brought down to room temperature and kept in desiccators for at least 15 min. The filter paper with the residue was again weighed, and the difference between the two weighed readings was taken for determining MLSS as follows:

$$\text{MLSS(mq/l)} = \frac{[(A - B) * 1000 * 1000]}{\text{Volume of sample}}$$

where

A is the sample and filter weight (gm)

B is the weight of the filter (gm).

3.10.3 MLVSS (Mixed Liquor Volatile Suspended Solids)

At first, weight of an empty crucible was taken, and simultaneously, muffle furnace was warmed up. The filter paper with the residue along with the crucible was kept in muffle furnace at 550 °C and kept for 20 min and weighed again. The difference between the two weighed readings was examined, and finally, the MLVSS value was calculated.

$$\text{MLVSS(mq/l)} = \frac{[E(\text{gm}) * 1000 * 1000]}{\text{Volume of sample(mL)}}$$

where $E = \{(A - B) - (C - D)\}$ (gm) = amount of volatile solids, where A is the sample and filter paper weight (gm)

B is the weight of the filter paper (gm)

C is the final weight of the crucible along with residue (gm)

D is the initial weight of the crucible (gm).

3.10.4 Alkalinity

Alkalinity is the acid neutralizing capacity and is equivalent to the sum of all titratable bases. Alkalinity values provide guidance in applying proper doses of chemicals in water and wastewater treatment processes, particularly in coagulation, softening and operational control of anaerobic digestion. The alkalinity of wastewater is primarily a function of carbonate, bicarbonate and hydroxide ions content and is represented as an indication of the concentrations of these ions. Titration was done with the bromocresol green indicator and was continued till the color of the sample sharp changes from blue to green which indicates the complete neutralization of the ions constituting the total alkalinity.

3.10.5 Total Solids

A well-mixed sample is evaporated in a weighed dish in an oven at 103–105 °C. The increase in weight over that of the empty dish represents total solids.

$$\text{Total solid content in mg/l} = \frac{(W_2 - W_1) * 10^6}{\text{Volume of sample in mL}}$$

W_1 = weight of empty dish in gm. W_2 = weight of dish with sample in gm.

3.10.6 Total Suspended Solids

A well-mixed sample is filtered through a weighed standard glass fiber filter, and the residue retained on the filter is dried to a constant temperature of 103°–105 °C. The increase in weight of the filter represents the TSS.

$$\text{Total solid content in mg/l} = \frac{(W_2 - W_1) * 10^6}{\text{Volume of sample in mL}}$$

W_1 = weight of filter paper before filtration in gm.

W_2 = weight of filter paper including filtrates after filtration in gm.

3.10.7 Determination of Dissolved Oxygen (DO)

Dissolved oxygen (DO) is measured by azide modification of Winkler's method based on iodometric principle. Oxygen oxidizes Mn^{2+} to a higher state of valency (+4) under alkaline conditions. If no oxygen is present, a white precipitate of $Mn(OH)_2$ is formed. In the presence of oxygen, Mn^{2+} is stoichiometrically oxidized to MnO_2, which forms brown hydrated oxide at high pH. This step is called fixation of oxygen. Addition of H_2SO_4 to sample in the presence of iodide results in the oxidized manganese (Mn^{4+}) to revert to the divalent state, with the liberation of iodine equivalent to the original DO content. Liberated iodine is titrated against N/40 sodium thiosulfate solution. Certain oxidizing agents such as nitrate and ferric ions are capable of oxidizing I to I_2 and thus interfere with the measurement. Nitrite interference can be overcome by introducing sodium azide, which reacts with nitrite under acidic conditions to form N_2, N_2O and H_2O.

For measurement of DO, remove the stopper and add 1 mL of $MnSO_4$ solution, followed by 1 mL of alkali-iodide-azide solution by inserting the pipette well below the surface of the liquid. Put the stopper with care to exclude air bubbles completely and mix by inverting the bottle several times. When precipitate has settled sufficiently to leave a clear supernatant above manganese hydroxide floc, add 1 mL conc. H_2SO_4. Stopper was put again and mixed by inverting several times until dissolution of the precipitate is complete and there is uniform distribution of liberated iodine. Titrate a volume (201 mL) corresponding to 200 mL of the original sample after correction for sample loss by displacement with reagents. Titration with standard N/40 $Na_2S_2O_3$ solution is done to a pale straw color, and then, a few drops of starch solution is added to proceed titration up to the first disappearance of the blue color. DO in mg/l is determined from the volume (mL) of the standard $Na_2S_2O_3$ solution consumed in titration.

3.10.8 Ammonium Nitrogen ($NH4^+$-N)

50 mL of sample or its suitable aliquot was placed in 800-mL Kjeldahl flask. The aliquot was diluted to 300 mL in the flask. 25 mL of borate buffer and 6(N) NaOH were added until the content in the flask is reduced to pH 9.5. By putting the glass beads, the sample solution in the flask was boiled. In the distillation process, a conical flask containing 50 mL of boric acid was kept at the receiving end. Now, the distillation of the sample solution was continued till the volume of the conical flask reached about 250 mL. Then, the content of the conical flask is titrated with N/50 conc. H_2SO_4 till the end point color changes from green to blue, and the ammonium nitrogen was calculated as follows:

$$\mathrm{NH}_4^+ - \mathrm{N(mg/l)} = \frac{(A - B) * 280}{\text{Volume of sample(mL)}}$$

where A = volume of H_2SO_4 titrated for sample,

B = volume of H_2SO_4 for blank.

3.10.9 Total Kjeldahl Nitrogen (TKN)

50 mL of sample or its suitable aliquot was placed in 800-mL Kjeldahl flask. Thereafter, 6.7 mL of conc. H_2SO_4, 6.7 g of K_2SO_4 and 0.364 g of CuSO4 was added to the flask. The sample solution was diluted to 100 mL and was digested in the heater till the volume is reduced to 25–50 mL. The 25–50 mL of sample was then diluted to 300 mL in the flask. 25 mL of borate buffer and 6(N) NaOH were added until the content in the flask is reduced to pH 9.5. By putting the glass beads, the sample solution in the flask was boiled. In the distillation process, a conical flask containing 50 mL of boric acid was kept at the receiving end. Now, the distillation of the sample solution was continued till the content volume of the conical flask reached about 250 mL. Then, the content of the conical flask was titrated with N/50 conc. H_2SO_4 till the end point color changes from green to blue, and the ammonium nitrogen was calculated as follows:

$$\mathrm{TKN(mg/l)} = \frac{(A - B) * 280}{\text{Volume of sample(mL)}}$$

where A = volume of H_2SO_4 titrated for sample,

B = volume of H_2SO_4 for blank.

3.10.10 Organic Nitrogen

Organic nitrogen was determined as the difference between total Kjeldahl nitrogen and ammonium nitrogen.

3.10.11 Total Phosphorus

Stannous chloride method (Ref. Standard Methods 1995) was followed for the determination of phosphorus in the wastewater sample. In a 50 mL sample, 2 mL of molybdate reagent (comprising of ammonium molybdate and conc. H_2SO_4) and 0.25 mL (5 drops) of stannous chloride reagent were added and properly mixed.

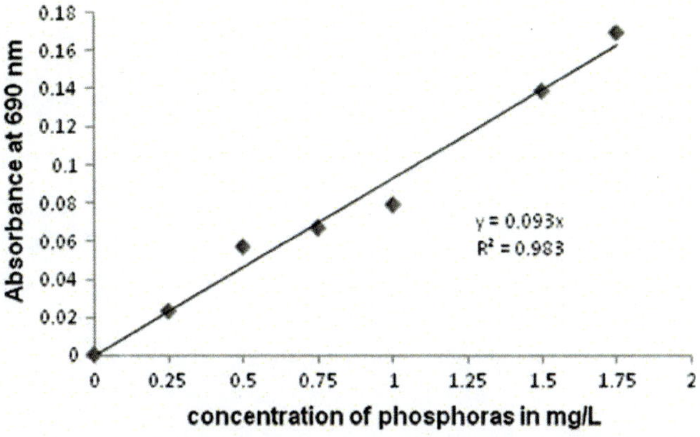

Fig. 3.8 Calibration graph for the determination of concentration of Phosphorus

After 10 min, but before 12 min, the color of the mixed sample was photometrically measured at 690 nm in spectrophotometer and the absorbance reading was compared with a calibration curve to get the value of phosphorus concentration. The calibration curve used for the determination of total phosphorus is shown in Fig. 3.8.

References

1. Cox HJ, et al (1999) Chemical removal of biomass from waste air biotrickling filters: Screening of chemicals of potential interest. Water Res 33:2383–2391
2. Lazarova V, Pierzo V, Fontvielle D, Manem J (1994) Integrated approach for biofilm characterization and biomass activity control. Water Sci Technol 29(7):345–354
3. Herbert D, Philips PJ, Strange JE (1971) Chemical analysis of microbial cell of the book (Chap. 3). In: Methods in microbiology, vol V, Part B, pp 209–344
4. Lowry OH, Rosebrough NJ, Fair AL, Randall RJ (1951) Protein measurement with Folin Phenol Reagent. J Biol Chem 193:265–275
5. Standard Methods for the examination of Water and Wastewater, prepared and published jointly by American Public Health Association, American Water Work Association and Water Pollution Control Federation; Sixteenth Edition (1995)

Chapter 4
Development of a Mathematical Model for an Aerobic Fixed-Bed Hybrid Bioreactor and Its Analytical Validation

4.1 Basic Consideration

In order to design a fixed-bed hybrid biological reactor, a rational and simplified model is very much useful. There are a limited number of methods available for solving problems on hybrid bioreactor system. In the hybrid biological process, solution of biofilm model is much tedious and cumbersome compared to the suspended growth. The reliable analysis for predictions pertinent to the performance of the said reactor is still unavailable. Most kinetic models assumed single growth, i.e., either suspended or attached in sequential fashion. Even where the competition for rate-limiting substrates between two growths (both suspended and attached) was simultaneously considered, no unique, accurate and simplified solution was derived from the steady-state substrate mass balance and biomass balance equations.

4.1.1 Assumptions for Model Development

Following assumptions have been made in the development of a mathematical model of a fixed-bed hybrid bioreactor.

- The model has been derived assuming simultaneous growth of both suspended and attached biomass in the reactor.
- A fraction of substrate is uniformly exposed to the suspended biomass. The remaining fraction of substrate flows through the biofilm–liquid interface and then through the biofilm.
- Monod kinetics is followed for the utilization of carbonaceous substrate assuming no inhibition.
- Hybrid bioreactor is considered as completely mixed, and thus, the entering substrate concentration S_0 becomes exiting substrate concentration S_w in the bulk liquid.

S. Sarkar and D. Mazumder, *Fixed Bed Hybrid Bioreactor*, Green Energy and Technology,
https://doi.org/10.1007/978-981-33-4546-1_4

- The model essentially considered the substrate mass transport into the biofilm as per Fick's law.
- Average substrate flux (J_{avg}) has been determined considering a variation in substrate flux from biofilm layer to layer on account of substrate gradient.
- To calculate the average value of "substrate flux" (J_{avg}) into the biofilm, 5 (five) divisions (based on equal interval of substrate concentration) inside the biofilm have been considered.
- As a rational approach of calculation of the overall substrate flux, an area-weighted average value of individual fluxes in various layers has been taken.
- J_0, the substrate flux corresponding to S_{min} at attachment surface, is considered zero.

4.1.2 Fundamental Concept

The mathematical model for fixed-bed hybrid bioreactor is based on steady-state mass balance of both carbonaceous substrate and biomass under suspended and attached growth simultaneously along with substrate mass transport into the biofilm. Monod kinetics is followed for the utilization of carbonaceous substrate assuming no inhibition. Average substrate flux on account of varying substrate concentration profile inside the biofilm is considered for determining the value of exiting substrate concentration (S_w), which is a key output variable for evaluating the performance of the reactor. Substrate concentration S_0 becomes exiting substrate concentration S_w in the bulk liquid due to complete mixing in the reactor. This concentration S_w is the homogeneous bulk liquid concentration up to the liquid–biofilm interface, which further decreases to S_{min}, i.e., the minimum substrate concentration at biofilm attachment surface. Firstly, the combination of mass balance equation of substrate under both the suspended and the attached growth is derived. In this combined mass balance equation of substrate, the expression of suspended biomass concentration is substituted by the term derived from the biomass balance equation of the suspended growth accounting for biomass shear loss from biofilm. Average substrate flux is computed from the solution of the equation of substrate mass balance into biofilm.

Finally, exiting substrate concentration (S_w) is obtained substituting the value of average substrate flux in the combined mass balance equation of substrate, which is initially derived. Applying the Runge–Kutta method of analysis of the substrate mass balance equation of biofilm and applying the boundary conditions, effective biofilm thickness (L_e) is obtained. L_e is determined by iteration process, where convergence of iteration is attributed at the liquid–biofilm interface when computed substrate concentration (S_w) becomes equal to the assumed one. The effective biofilm thickness contains those biomasses actively metabolizing the substrate. Beyond this effective biofilm thickness, the substrate flux ceases to get utilized further. In fact, by the effective biofilm thickness of biofilm, it can be ascertained whether the biofilm is deep or shallow [1].

4.1.3 *Justification of Current Methodology*

The present mathematical model employed the kinetics of suspended and attached-growth biomass in integrated manner considering their simultaneous growth. Monod kinetics is followed for the utilization of carbonaceous substrate considering no inhibition in the application of this mathematical model for the treatment of municipal wastewater.

J_{avg} (average substrate flux) is considered in mathematical model, because there is always a variation in substrate flux from biofilm layer to layer on account of substrate gradient. The substrate gradient is exerted within individual biofilm layer, when the substrate incoming due to diffusion gets utilized by the respective biofilm layer. In fact, the substrate gradient becomes a driving force to allow for the movement of the substrate through individual biofilm layer. Thus, the substrate concentration gradually decreases from S_w at the biofilm–liquid interface to S_{min} at the attachment surface.

From the definition of substrate flux (J) which essentially equals $\int_0^{lf}(-\text{rut})dz$ (Rittmann and Mccarty 2001), the overall utilization of substrate throughout the biofilm needs to be taken into account. Therefore, as a rational approach of calculation of the overall substrate flux, an area-weighted average value of individual fluxes in various layers has been taken.

The accuracy of result for calculating the average substrate flux would be better if the biofilm thickness is divided into several layers (since it is not a linear curve). Therefore, calculations were made for evaluating the J_{avg}, with 3, 4, 5, 6, 7 and 8 layers in the biofilm. Interestingly, the values of effluent substrate concentration (S_w) as well as substrate flux (J) were converging for five (5) layers, and thereafter, there was hardly any deviation of results when the number of layers increased to five (5) and onward. In view of that, five (5) layers have been considered in the mathematical model. In this mathematical model, efforts have been put for removing of carbonaceous organic matter in the treatment of municipal wastewater. Thus, the output variable S_w in the model actually indicates the effluent substrate concentration in the treatment of wastewater.

The output variable, effective biofilm thickness (L_e), is also considered in the model as it contains those biomasses actively metabolizing the substrate beyond which the substrate flux ceases to get utilized further. Runge–Kutta numerical analysis method is found most appropriate for determining the effective biofilm thickness and is used for the solution of substrate mass balance equation in the biofilm.

4.2 Development of a Mathematical Model

A mathematical model for hybrid bioreactor is developed with a computer programming in FORTRAN and excel worksheet to easily calculate the output parameters like exiting substrate concentration, substrate flux, effective and total biofilm thickness. A simplified and functionally efficient model helps in the process design of a hybrid bioreactor. One of the targets of the present research work is to derive a process design modality in respect of aerobic hybrid biological reactor. The model is also developed with an objective to desire a methodology for determining the kinetic coefficients of the hybrid bioreactor.

4.2.1 Problem Scenario

There is no process design model so far for a hybrid bioreactor system without which it is difficult to build up such a reactor system for the treatment of wastewater. In order to go for a suitable process design, a simplified mathematical model is required for determining the outputs rationally. Obviously, all the outputs are finally required for finalizing the design of an aerobic hybrid bioreactor. The mathematical model of a hybrid bioreactor is an integration of the biofilm model and the classical suspended-growth model. Out of these two models, suspended-growth model is almost unique and simple which is based on Monod's growth kinetics. But, the analysis of biofilm model appears to be cumbersome on account of its complexity. Besides, the characteristic of biofilm uptake is quite different from the suspended-growth biomass. Although there are many existing models on biofilm, the procedure of solution is very tedious and the results of outputs are found approximate. There are a limited number of modelings in the integrated biofilm and suspended-growth process which is required for the design of hybrid bioreactor. Thus, the necessity of a model for hybrid bioreactor can be reasonably thought for the process design of the reactor.

4.2.2 Model Description

The schematic diagram of fixed-bed hybrid bioreactor is shown in Fig. 4.1.

The steady-state substrate balance for suspended growth is

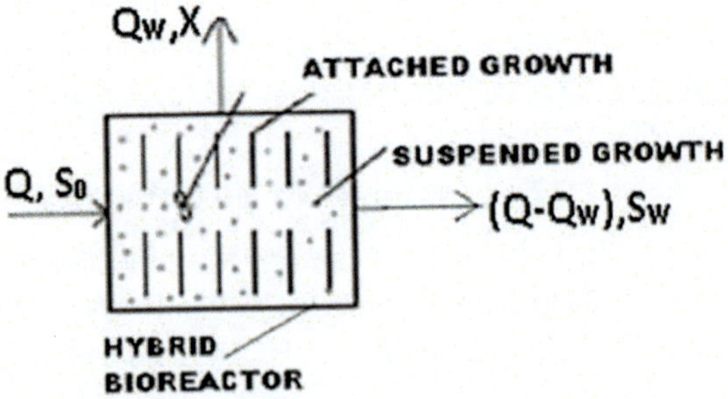

Fig. 4.1 Schematic diagram of a hybrid bioreactor

$$r[Q(S_0 - S_w)]\frac{pVkXS_w}{K + S_w} = 0 \tag{4.1}$$

where

S_0 = entering substrate concentration (mg/cm^3).

S_w = exiting substrate concentration (mg/cm^3) in the bulk liquid. Q = inflow rate (m^3/s), r = fraction of substrate used by suspended growth, p = porosity of hybrid reactor, V = empty-bed volume of hybrid reactor (m^3), k = maximum specific rate of substrate utilization (per day), X = concentration of suspended biomass in hybrid reactor (mg/cm^3), K = half-velocity coefficient (mg/cm^3).

Dividing Eq. (4.1) by Q, the following equation is obtained:

$$r[(S_0 - S_w)]\frac{p\theta kXS_w}{K + S_w} = 0 \tag{4.2}$$

where θ = empty-bed hydraulic detention time (h) = $\frac{V}{Q}$.

Now, the steady-state substrate balance for the attached growth is,

$$(1 - r)[Q(S_0 - S_w)] - aJ_{avg}V = 0 \tag{4.3}$$

where a = specific surface area of supporting media (cm^{-1}), J_{avg} = average substrate flux into the biofilm (mg/cm^2/day). Again, dividing Eq. (4.3) by Q, the following equation is obtained:

$$(1 - r)(S_0 - S_w) - aJ_{avg}\theta = 0 \tag{4.4}$$

Now, combining the two substrate balance Eqs. (4.2) and (4.4) for both the suspended growth and attached growth the following equation is obtained:

$$S_0 - S_w - \frac{pkXS_w\theta}{K + S_w} - aJ_{avg}\theta = 0. \tag{4.5}$$

The substrate profile within the biofilm accounting for its utilization by the attached biomass can be represented in Fig. 4.2.

To calculate the average value of "substrate flux" (J_{avg}) into the biofilm, 5 (five) divisions (*based on equal interval of substrate concentration*) inside the biofilm have been considered as shown in Fig. 4.3.

Such an arrangement of divisions is conceptualized within the biofilm for determining the accurate value of individual substrate flux at respective division. It has also been observed that there is hardly any deviation of results in case number of divisions more than five (Sarkar and Mazumder 2016).

Now, $J_{avg} = (J_0 + J_1 + J_2 + J_3 + J_4 + J_5)/6$ [2], where J_1, J_2, J_3, J_4 and J_5 are substrate fluxes corresponding to substrate concentrations S_1, S_2, S_3, S_4 and S_w,

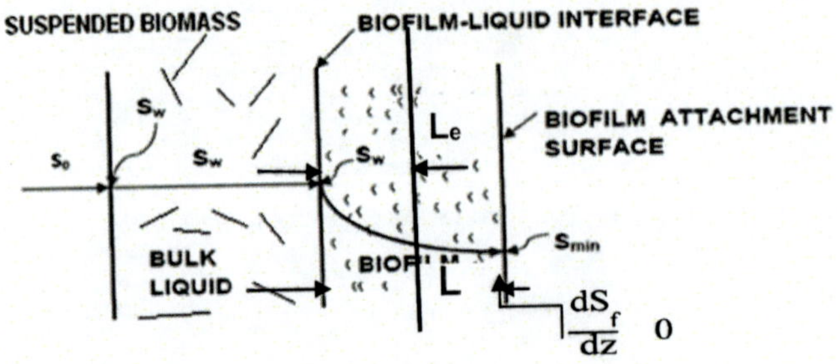

Fig. 4.2 Profile of substrate concentration in a hybrid bioreactor

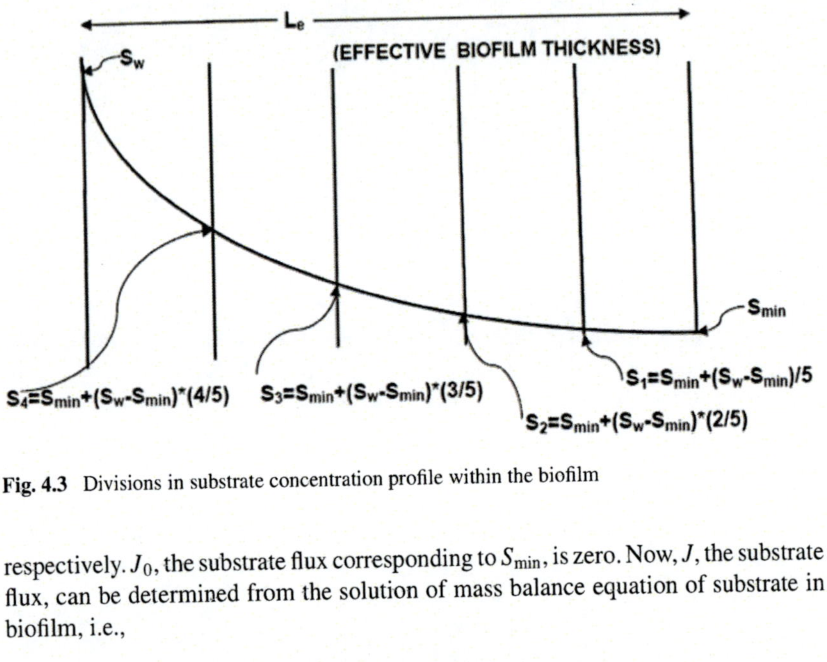

Fig. 4.3 Divisions in substrate concentration profile within the biofilm

respectively. J_0, the substrate flux corresponding to $S_{\min}$, is zero. Now, J, the substrate flux, can be determined from the solution of mass balance equation of substrate in biofilm, i.e.,

$$\frac{d^2 S_f}{dz^2} = \frac{k X_f S_f}{D_f(K + S_f)} \tag{4.6}$$

Multiplying both sides of Eq. (4.6) by $2\left(\frac{dS_f}{dz}\right)$ and integrating both sides, we get,

$$\left(\frac{dS_f}{dz}\right)^2 = 2kX_f \int \frac{S_f}{D_f(K + S_f)} dS_f$$

$$\Rightarrow \left(\frac{dS_f}{dz}\right)^2 = \frac{2kX_f}{D_f} \int 1 - \left(\frac{K}{(K + S_f)}\right) dS_f$$

$$\Rightarrow \left(\frac{dS_f}{dz}\right)^2 = \frac{2kX_f}{D_f}\left[S_f - K\log(K + S_f)\right] + A$$

Now, when $S_f = S_w$, $\frac{dS_f}{dz} = 0$

$$\Rightarrow 0 = \frac{2kX_f}{D_f}[S_w - K\log(K + S_w)] + A$$

$$A = -\frac{2kX_f}{D_f}[S_w - K\log(K + S_w)]$$

$$\Rightarrow \left(\frac{dS_f}{dz}\right)^2 = \frac{2kX_f}{D_f}[S_f - K\log(K + S_f)] - \frac{2kX_f}{D_f}[S_W - K\log(K + S_W)]$$

$$\Rightarrow \left(\frac{dS_f}{dz}\right)^2 = \frac{2kX_f}{D_f}\left[(S_f - S_W) + K\log\frac{(K + S_W)}{(K + S_f)}\right]$$

$$\Rightarrow \left(\frac{dS_f}{dz}\right) = \sqrt{\frac{2kX_f}{D_f}\left[(S_f - S_W) + K\log\frac{(K + S_W)}{(K + S_f)}\right]}$$

Hence,

$$J = \left(\frac{dS_f}{dz}\right)D_f = D_f\sqrt{\frac{2kX_f}{D_f}\left[(S_f - S_W) + K\log\frac{(K + S_W)}{(K + S_f)}\right]}$$

$$J = \sqrt{2kX_fD_f\left[(S_{entry} - S_{exit}) + K\log\frac{(K + S_{exit})}{(K + S_{entry})}\right]} \tag{4.7}$$

where S_f = substrate concentration at any point in the biofilm (mg/cm^3),
X_f = active biomass density within the biofilm(mg/cm^3),

D_f = molecular diffusion coefficient of the substrate in the biofilm (cm^2/day),

S_{entry} = entering substrate concentration in a small segment in the biofilm (mg/cm^3),

S_{exit} = exiting substrate concentration in a small segment in the biofilm (mg/cm^3),

$$S_{min} = K^* \frac{b_e}{Y^*k - b_t},$$

where S_{min} = minimum concentration of rate-limiting substrate at biofilm attachment surface (mg/cm^3).

Equation (4.7) can be solved to find out 'J' at different points of substrate concentrations as follows:

$$J_1 = \sqrt{2kX_f D_f[(S_1 - S_{min}) + K\ln[(K + S_{min})/(K + S_1)]} \tag{4.8}$$

$$J_2 = \sqrt{2kX_f D_f[(S_2 - S_1) + K\ln[(K + S_1)/(K + S_2)] + (J_1)^2} \tag{4.9}$$

$$J_3 = \sqrt{2kX_f D_f[(S_3 - S_2) + K\ln[(K + S_2)/(K + S_3)] + (J_2)^2} \tag{4.10}$$

$$J_4 = \sqrt{2kX_f D_f[(S_4 - S_3) + K\ln[(K + S_3)/(K + S_4)] + (J_3)^2} \tag{4.11}$$

$$J_5 = \sqrt{2kX_f D_f[(S_w - S_4) + K\ln[(K + S_4/(K + S_w)] + (J_4)^2} \tag{4.12}$$

where J_1, J_2, J_3, J_4 and J_5 are substrate fluxes corresponding to substrate concentrations S_1, S_2, S_3, S_4 and S_w, respectively. It is obvious that the substrate flux (say J_0) corresponding to S_{min} is zero. Therefore, the arithmetic average of J can be calculated as

$$J_{avg} = (J_0 + J_1 + J_2 + J_3 + J_4 + J_5)/6. \tag{4.13}$$

Now, from the steady-state mass balance of active microorganisms in a biofilm, as well as under suspended-growth state (Referring to the conceptual diagram of hybrid bioreactor with recirculation as shown in Fig. 5d, Annexure) II-b, we get,

$$YaJ_{avg}V\frac{b_s}{b_t} - Q_wX - (Q - Q_w)X_e + pVX\left(\frac{YKS_w}{K + S_w} - b_d\right) = 0 \tag{4.14}$$

In Eq. (4.8), shearing losses from the biofilm, i.e., $YaJ_{avg}V\frac{b_s}{b_t}$, also enter the suspended-growth regime.

Now, dividing Eq. (4.8) by pVX,

$$\frac{YaJ_{avg}}{pX}\frac{b_s}{b_t} - \frac{1}{\theta_c} + \left(\frac{YkS_w}{K + S_w} - b_d\right) = 0,$$

where, $\theta_c = \frac{pVX}{Q_wX + (Q - Q_w)X_e}$

$$\text{i.e., } X = \frac{YaJ_{avg}\frac{b_s}{pb_t}}{\frac{1}{\theta_c} + b_d - \frac{YkS_w}{K + S_w}}, \tag{4.15}$$

where,

Y = bacteria yield coefficient,

b_s = biomass loss rate due to shearing from biofilm, day^{-1},

b_t = total biomass loss rate from biofilm, day^{-1},

b_d = biomass decay coefficient, day^{-1},

θ_c = mean cell residence time or solid retention time (day).

Putting the value of X in Eq. (4.5), we get,

$$S_0 - S_w - \frac{kS_w\theta Ya J_{avg}\frac{b_s}{b_t}}{(K + S_w)\left(\frac{1}{\theta_c} + b_d - \frac{YkS_w}{K+S_w}\right)} - aJ_{avg}\theta = 0 \tag{4.16}$$

Equation (4.16) can be reformed as,

$$S_0 - S_W - A1J_{avg} - aJ_{avg}\theta$$

$$\left[\text{considering } A1 = \frac{kS_w\theta Ya\frac{b_s}{b_t}}{(K+S_w)\left(\frac{1}{\theta_c}+b_d-\frac{YkS_w}{K+S_w}\right)}\right]$$

i.e.,

$$S_0 - S_w = J_{avg}(A1 + a\theta) \tag{4.17}$$

Now, from Eqs. (4.7) and (4.17), by process of iteration in a computer program (FORTRAN), S_w, the exiting substrate concentration in the bulk liquid can be determined.

Now, for calculating the effective biofilm thickness L_e, Runge–Kutta method can be applied in the equation $\frac{d^2S_f}{dz^2} = \frac{kX_fS_f}{D_f(K+S_f)}$ for numerical solution as follows:

$$\frac{d^2S_f}{dz^2} = f\left(z, \frac{dS_f}{dz}\right), \quad \frac{dS_f}{dz}(z_0) = \frac{dS_{f0}}{dz} = K1$$
$$= 0 \;\; [S_{f0} = S_{min} \text{ at } z = 0]$$

$$L_1 = f\left(z_0, \frac{dS_{f0}}{dz}\right) = \frac{d^2S_{f0}}{dz^2} = \frac{kX_fS_{f0}}{D_f(K + S_{f1})}$$

$$\frac{dS_{f1}}{dz} = \frac{dS_{f0}}{dz} + 0.5 * L1 * h = K2,$$

where h = step = effective biofilm thickness (cm).

h is the distance between $z = 0$ (at the attachment surface) and $z = L_e$ (at the biofilm–water interface)

$$L2 = h * f\left(z_0 + \frac{h}{2}, \frac{dS_n}{dz}\right) = h * \frac{d^2 S_{f1}}{dz^2} = \frac{kX_f(S_{f0} + 0.5K1 * h)}{D_f(K + S_{f0} + 0.5K1 * h)}$$

$$\frac{dS_{f2}}{dz} = \frac{dS_{f0}}{dz} + 0.5 * L2 * h = K3$$

$$L3 = h * f\left(z_0 + \frac{h}{2}, \frac{dS_{f2}}{dz}\right) = h * \frac{d^2 S_{f2}}{dz^2} = \frac{kX_f(S_{f0} + 0.5K2 * h)}{D_f\,(K + S_{f0} + 0.5K2 * h)}$$

$$\frac{dS_B}{dz} = \frac{dS_{f0}}{dz} + L3 * h = K4$$

$$L4 = h * f\left(Z_0 + h, \frac{dS_{f3}}{dz}\right) = h^* \frac{d^2 S_{f3}}{dz^2} = \frac{kX_f\left(S_{f0} + K3^* h\right)}{D_f\left(K + S_{f0} + K3^* h\right)}$$

$$\Delta Y(1) = \frac{h}{6} * (K1 + 2 * K2 + 2 * K3 + K4) \tag{4.18}$$

$$\Delta Y(2) = \frac{h}{6} * \left(L1 + 2 * L2 + 2 * L^3 + L4\right) \tag{4.19}$$

Here, $\Delta Y(1)$ stands for increment of substrate concentration, and $\Delta Y(2)$ stands for $\frac{dS_f}{dz}$.

Therefore,

$$S_w = S_{min} + \Delta Y(1) \tag{4.20}$$

$$J_5 = D_f * \Delta Y(2) \tag{4.21}$$

$$L_f = \frac{J * Y}{X_f * b_t}, \text{ where } L_f = \text{total biofilm thickness(cm)} \tag{4.22}$$

In order to calculate S_w, J and L_e using Eqs. (4.5)–(4.21), i.e., to solve the hybrid bioreactor model, two flowcharts were constructed as shown in Fig. 4.4 and Fig. 4.5.

Consequently, two detailed FORTRAN programs as shown in Program 1 and Program 2 below have been developed on the basis of those flowcharts.

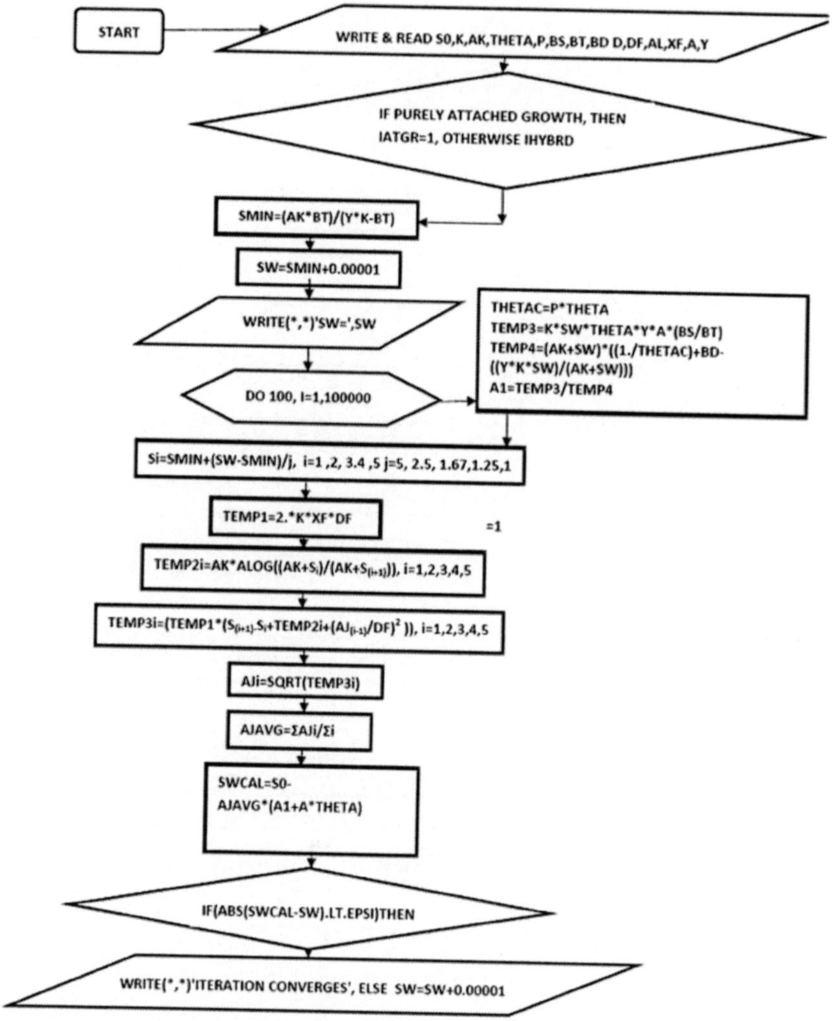

Fig. 4.4 Flowchart for programming in FORTRAN to calculate unknown exiting substrate concentration S_w and average substrate flux J_{avg}

Program 1 For finding out S_w and J_{avg}

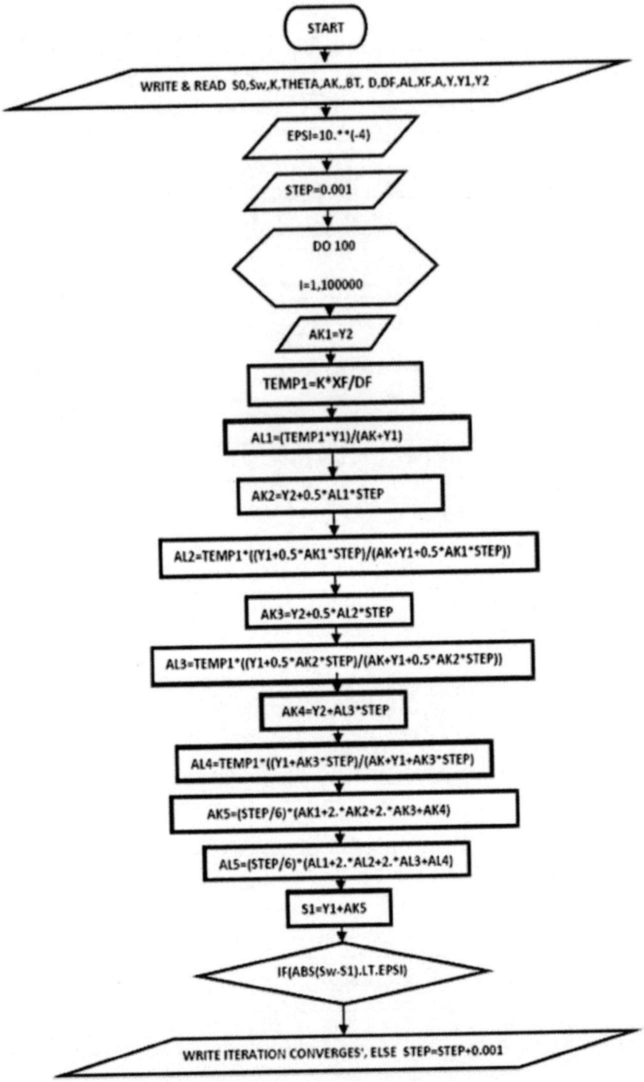

Fig. 4.5 Flow Chart for programming in FORTRAN to calculate Effective Biofilm Thickness (L_e) under a known Exiting Substrate Concentration (S_w)

```
 PROGRAMME 1 ANALYSIS FOR HYBRID BIOREACTOR
 WRITE(*,*)'INPUT THE VALUES OF S0,BK,AK,THETA,P,BS,BT,BD, D,DF,AL,XF,A,Y'
 READ(*,*)S0,BK,AK,THETA,P,BS,BT,BD, D,DF,AL,XF,A,Y
 THETA=THETA/24
 WRITE(*,*)'IF PURELY ATTACHED GROWTH, THEN IATGR=1'
 WRITE(*,*)'IF BOTH SUSPENDED AND ATTACHED GROWTH, THEN IHYBRD=1'
 WRITE(*,*)'INPUT IATGR,IHYBRD'
 READ(*,*)IATGR,IHYBRD
 EPSI=10.**(-3)
 SMIN=(AK*BT)/(Y*BK-BT)
 SW=SMIN+0.0001
10 SW=SW
 WRITE(*,*)'SW=',SW
 KOUNT=100000
 DO I=1,KOUNT
 CONTINUE
 THETAC=P*THETA
 TEMP3=BK*SW*THETA*Y*A*(BS/BT)
 TEMP4=(AK+SW)*((1./THETAC)+BD-((Y*BK*SW)/(AK+SW)))
 A1=TEMP3/TEMP4
 IF (IATGR.EQ.1) THEN
 A1=0
 ELSE
 END IF
 S1=SMIN+0.2*(SW-SMIN)
 TEMP1=2.*BK*XF*DF
 TEMP21=AK*ALOG((AK+SMIN)/(AK+S1))
 TEMP31=(TEMP1*(S1-SMIN+TEMP21))
 AJ1=SQRT(TEMP31)
 S2=SMIN+0.4*(SW-SMIN)
 TEMP22=AK*ALOG((AK+S1)/(AK+S2))
 TEMP32=(TEMP1*(S2-S1+TEMP22))+((AJ1)*(AJ1))
 AJ2=SQRT(TEMP32)
 S3=SMIN+0.6*(SW-SMIN)
 TEMP23=AK*ALOG((AK+S2)/(AK+S3))
 TEMP33=(TEMP1*(S3-S2+TEMP23))+((AJ2)*(AJ2))
 AJ3=SQRT(TEMP33)
 S4=SMIN+0.8*(SW-SMIN)
 TEMP24=AK*ALOG((AK+S3)/(AK+S4))
 TEMP34=(TEMP1*(S4-S3+TEMP24))+((AJ3)*(AJ3))
 AJ4=SQRT(TEMP34)
 S5=SW
 TEMP25=AK*ALOG((AK+S4)/(AK+S5))
 TEMP35=(TEMP1*(S5-S4+TEMP25))+((AJ4)*(AJ4))
 AJ5=SQRT(TEMP35)
 AJ0=0
 AJAVG=(AJ0+AJ1+AJ2+AJ3+AJ4+AJ5)/6
 SWCAL=S0-AJAVG*(A1+A*THETA)
 WRITE(*,*)'SWCAL=',SWCAL,'AJAVG=',AJAVG
 IF(ABS(SWCAL-SW).LT.EPSI)THEN
 WRITE(*,*)'ITERATION CONVERGES'
 WRITE(*,*)'SW=',SW,'AJAVG=',AJAVG
 STOP 'NORMAL END'
 ELSE
 SW=SW+0.00001
 GO TO 10
 END IF
 KOUNTC=MOD(I,100)
 IF (KOUNTC.EQ.1)THEN
 WRITE(*,*)'ITERATION DOES NOT CONVERGE'
 WRITE(*,*)'I=',I,'SW=',SW
 END IF
 IF(I.GE.KOUNT)THEN
 EXIT
 END IF
 END DO
 STOP'ABNORMAL END'
 END
```

Program 2 For finding out L_e

```
PROGRAMME 2 ANALYSIS FOR HYBRID BIORECTOR
WRITE(*,*)'INPUT THE VALUES OF S0,BK,AK,THETA,BT'
READ(*,*)S0,BK,AK,THETA,BT
THETA=THETA/24
WRITE(*,*)'INPUT THE VALUES OF D,DF,AL,XF,A,Y,Y2'
READ(*,*)D,DF,AL,XF,A,Y,Y2
EPSI=10.**(-3)
SMIN=(AK*BT)/(Y*BK-BT)
SW=SW
STEP=0.0001
05 STEP=STEP
WRITE(*,*)'SW=',SW,'STEP=',STEP
KOUNT=100000
DO I=1,KOUNT
TEMP1=2.*BK*XF*DF
TEMP2=AK*ALOG((AK+SMIN)/(AK+SW))
TEMP5=(TEMP1*(SW-SMIN+TEMP2))
AJ1=SQRT(TEMP5)
TEMP6=(BK*XF)/DF
Y1=SMIN
AK1=Y2
AL1=(TEMP6*Y1)/(AK+Y1)
AK2=Y2+0.5*AL1*STEP
AL2=TEMP6*((Y1+0.5*AK1*STEP)/(AK+Y1+0.5*AK1*STEP))
AK3=Y2+0.5*AL2*STEP
AL3=TEMP6*((Y1+0.5*AK2*STEP)/(AK+Y1+0.5*AK2*STEP))
AK4=Y2+AL3*STEP
AL4=TEMP6*((Y1+AK3*STEP)/(AK+Y1+AK3*STEP))
AK5=(STEP/6)*(AK1+2.*AK2+2.*AK3+AK4)
AL5=(STEP/6)*(AL1+2.*AL2+2.*AL3+AL4)
S1=Y1+AK5
AJ2=DF*(Y2+AL5)
IF(ABS(SW-S1).LT.EPSI)THEN
WRITE(*,*)'ITERATION CONVERGES'
WRITE(*,*) 'STEP=',STEP
STOP 'NORMAL END'
ELSE
STEP=STEP+0.0001
GO TO 5
END IF
KOUNTC=MOD(I,100)
IF (KOUNTC.EQ.1)THEN
WRITE(*,*)'ITERATION DOES NOT CONVERGE'
WRITE(*,*)'I=',I,'STEP=',STEP
END IF
IF(I.GE.KOUNT)THEN
EXIT
END IF
END DO
STOP'ABNORMAL END'
END
```

4.2.3 *Essence of Flowcharts Constructed*

Two flowcharts for computer programming in FORTRAN have been prepared approaching the iteration processes, with a view to calculate unknown exiting substrate concentration S_w, average substrate flux J_{avg} (Fig. 4.4) and effective biofilm thickness L_e. In the first flowchart (Fig. 4.5), equation numbers (4.7), (4.16) and (4.17) as stated earlier are simultaneously iterated to calculate the unknown exiting, i.e., bulk liquid substrate concentration S_w and unknown flux J. The initial iteration value S_w was assumed in this flowchart as any value higher than S_{min} (which is required to sustain a steady biofilm growth). The second flowchart (Fig. 4.5) utilized the value of S_w as obtained from the first flowchart to calculate L_e following the concept of Runge–Kutta method of analysis. The convergence of iteration is attributed at the liquid–biofilm interface when computed substrate concentration (S_w) becomes equal to the assumed one.

4.2.4 *Modality of Application of the Developed Model*

The solution method derived from the proposed model can be applied to determine the exiting substrate concentration, i.e., bulk liquid substrate concentration (S_w) and the average substrate flux (J_{avg}) by running the FORTRAN program (Program 1) based on the flowchart as shown in Fig. 4.4. After the determination of the values of S_w and J_{avg}, the FORTRAN program (Program 2) based on the flowchart as shown in Fig. 4.5 can be run for evaluating both the total and effective biofilm thicknesses L_f and L_e. The sequential steps of algorithm for Program 1 are stated in the Excel Worksheet as shown in Fig. 4.6.

Similarly, the sequential steps of algorithm for Program 2 are also stated in the Excel Worksheet as shown in Fig. 4.7.

4.2.5 *Novelty of the Developed Model*

In this mathematical model, simultaneous utilization of the substrate by suspended and attached biomass in a competitive manner is considered, which is very much essential for the process design of a fixed-bed hybrid bioreactor. A simplified model (with computer programming and Excel Worksheet) for hybrid bioreactor finds its importance for predicting the reliable outputs for the sake of process design. There is a flexibility of the present model making it a versatile one to find out the exiting substrate concentration along with substrate flux both in hybrid bioreactor and in a completely mixed biofilm reactor. No approximation is considered in this mathematical model depicting its uniqueness compared with the existing methods of numerical analysis. To maintain the steady-state biofilm model, it is necessary to check whether the value of effluent substrate concentration is more than minimum substrate concentration, i.e., S_{min} or not. This has been taken into consideration while developing the mathematical modeling of the hybrid bioreactor.

Apart from that, the developed model determines the substrate flux at any layer of the biofilm matrix along with the effective and total biofilm thickness. The model can determine the effective biofilm thickness which contains the effective biomass actively metabolizing the substrate beyond which the substrate flux ceases to get utilized further. In fact, by the effective thickness of biofilm, it can be ascertained whether the biofilm is deep or shallow. The application of Runge–Kutta method in this mathematical model is found, the only numerical analysis method by which the

MATHEMATICAL SOLUTION OF A HYBRID BIOREACTOR MODEL WHEN Sw NEEDS TO BE DETERMINED (A 11 STEPS ALGEBRAIC SOLUTION TO CALCULATE Sw)		
STEPS		
1	Sw(assumed)	$(K^*b_t)/(Yk-b_t)$+some value for initial trial
2	Si	$S_{min}+j^*(S_W-S_{MIN})$, i=1,2,3,4,5 and j=0.2, 0.4,0.6,0.8,1.0
3	GAMMA1	$2^*k^*X_f^*D_f$
4	GAMMA2	$K^*\ln((K+S_{i+1})/(K+S_i))$, i=1,2,3,4,5
5	GAMMA3	GAMMA1*$(S_{i+1}-S_i$+GAMMA2) +(J_m)^2, m=0,1,2,3,4,5, $J_0=0$
6	J_m	SQRT(GAMMA3)
7	Javg	SUM(J_m)/6
8	GAMMA4	$kS_w\Theta Ya(b_s/b_t)$
9	GAMMA5	(K+Sw)[1/ϴc+bd-(YkSw)/(K+Sw)]
10	A1	GAMMA4 / GAMMA5
11	Sw(Calculated)	S_0-Javg*(A1+aϴ/24)

WORKING SHEET FOR SOLUTION OF ABOVE STEPS				
	GIVEN PARAMETR	VALUE	GIVEN PARAMETR	VALUE
	S_0(Initial substrate conc(mg/cc)		b_t(total biomass loss rate , per day)	
	k(maxm.sp.rate of substrate use, per day)		b_s(biomass shear loss, per day)	
	ϴ(HRT,hr)		b_d(biomass decay per day)	
	Y(bacteria yield coefficient)		D(Molecular diffuisivity in liquid, sq cm/day)	
	a(sp. Surface area, per cm)		L (thickness of lquid layer,cm)	
	K(half velocity constant, mg/cc)		D_f (Molecular diffuisivity in biofilm, sq cm /day)	
			X_f (biomass conc. In biofilm, mg/cc)	
STEP 1		Sw(assumed)	****	
STEP 2		Si		
STEP 3		GAMMA1		
STEP 4		GAMMA2		
STEP 5		GAMMA3		
STEP 6		GAMMA4		
STEP 7		GAMMA5		
STEP 8		Javg(Calculated)		
STEP 9		Sw(calculated)		
****	S_w(actual) to be selected on trial and error till S_w(assumed) equals to S_w(calculated)			

Figure 4.6 Excel Worksheet to calculate unknown exiting substrate concentration S_w and average substrate flux J_{avg}

effective biofilm thickness can be determined in a process of iteration. The steady-state suspended biomass in the hybrid bioreactor can also be determined using the present mathematical model, which is essential for subsequent sludge management strategy. The present model is found very simple, fast and accurate in determining the output parameter required for the process design of hybrid bioreactor. No nomograph, normalized loading curve, etc., have been used in this model, and hence, there is no

MATHEMATICAL SOLUTION OF A HYBRID BIOREACTOR MODEL FOR CALCULATING EFFECTIVE BIOFILM THICKNESS Le(A 8 STEPS ALGEBRAIC SOLUTION TO CALCULATE L_e)		
STEPS		
1	Le (assumed)(cm)	A SMALL VALUE MAY BE TAKEN
2	L1	((k*X_f)/D_f)*Y1)/(K+Y1)
3	K2	Y2+0.5*L1*Le
4	L2	(((k*Xf)/Df)*(Y1+0.5*K1*Le))/(K+Y1+0.5*K1*Le)
5	K3	Y2+0.5*L2*Le
6	L3	(((k*X_f)/D_f)*(Y1+0.5*K2*Le))/(K+Y1+0.5*K2*Le)
7	K4	Y2+L3*Le
8	Le (CALCULATED)	((Sw.Y1)/(K1+2*K2+2*K3+K4))*6

WORKING SHEET FOR SOLUTION OF ABOVE STEPS				
	GIVEN PARAMETR	VALUE	GIVEN PARAMETR	VALUE
	S_0(Initial substrate conc(mg/cc)		b_t(total biomass loss rate , per day)	
	k(maxm.sp.rate of substrate use, per day)		D(Molecular diffuisivity in liquid, sq cm/day)	
	Θ(HRT,hr)		D_f(Molecular diffuisivity in biofilm, sq cm /day)	
	Y(bacteria yield coefficient)		L(thickness of liquid layer,cm)	
	a(sp. Surface area, per cm)		X_f(biomass conc. In biofilm, mg/cc)	
	K(half velocity constant, mg/cc)		Y2=dS_0/dZ at Z=0	
	K1=Y2 ie dS/dZ at Z=0, ie at attachment surface.		Y1=S_{min}(mg/cc)	
STEP 1		Le (assumed)	****	
STEP 2		L1		
STEP 3		K2		
STEP 4		L2		
STEP 5		K3		
STEP 6		L3		
STEP 7		K4		
STEP 8		L_e (CALCULATED)		
****	L_e needs to be selected on trial and error till L_e (assumed) equals to L_e(Calculated)			

Fig. 4.7 Excel worksheet to calculate Effective Biofilm Thickness (L_e) under a known Exiting Substrate Concentration (S_w)

chance of any manual error. Considering all the aspects, the present solution method can be proposed for rational design of the fixed-bed hybrid bioreactor with a good accuracy.

Table 4.1 Input data sets used for analytical validation of the mathematical model

Data set no.	Input substrate concentration ($S0$) (mg/cc)	HRT (θ) (h)	SRT (θ_c) (h)	Specific surface area (a) (cm^{-1})	Attached biomass density (Xf) (mg/cc)
1	0.43	24	12	0.9	25
2	0.43	12	12	1.8	25
3	0.16	12	12	0.154	54.63

4.3 Analytical Validation of the Developed Model

In this section, primary approach for analytical validation is firstly stated. Selected input data used for analytical validation are also highlighted in tabular form. Finally, modality of validation has been explained in detail to get an idea about the procedure of analytical validation of the present mathematical model.

4.3.1 Primary Approach

Firstly, the mathematical models for fixed-bed hybrid bioreactor have been identified as per the existing literatures. Then, the model outputs have been determined with a set of input data for all the existing models and the present model is developed. The present model is run with computer programs, Program 1 and Program 2, which are formulated as per the flow sheets. Finally, the outputs of present mathematical model are compared with those of existing models for same input data set in tabular form.

4.3.2 Selection of Data Set

Three data sets selected from the existing literatures have been used for analytical validation of the mathematical model which are stated in Table 4.1.

4.3.3 Modality of Validation

The present model for the hybrid bioreactor is used to determine the relevant output parameters analytically under the different sets of input parameters. The effluent substrate concentration (S_w), the substrate flux (J), the effective biofilm thickness (L_e) and the total biofilm thickness (L_f) are considered as the output parameters. The analytical solution by the present model is designated as Case 1. The same analytical solution is compared with the solution using different existing methods as demonstrated in available literatures to judge the efficacy of the present model. The model analysis by [3], Fouad and Bhargava (2004) and [4] is denoted as Case 2, Case 3 and Case 4, respectively. Table 4b of Annexure Ia illustrates the comparison of output parameters between Case 1 and Cases 2, 3 and 4 separately. In the comparison analysis, all necessary kinetic coefficients and input parameters like initial substrate concentration (S_0), hydraulic retention time (θ), sludge retention time (θ_c), specific surface area (a) and attached biomass density (X_f) are chosen as per the standard literatures of the respective cases. The results of analysis as shown in Table-5b of Annexure II-a depict the outcomes from computer programming of the present model which are compared with the existing models.

The kinetic coefficients and physical data for comparison between Case 1 and Case 2, and Case 1 and Case 3 are considered as: k = 10 day-1, $Y = 0.45$, $K = 0.01$ mg/cm^3, $b_t = 0.41$ day^{-1}, $b_s = 0.21$ day^{-1} and $b_d = 0.2$ day^{-1} $D = 1.25$ cm^2/day, $D_f = 0.75$ cm^2/day, $L = 0.0078$ cm. The kinetic coefficients and physical data for comparison between Case 1 and Case 4 is considered as: $k = 2.08$ day^{-1}, $Y = 0.72$, $K = 0.087$ mg/cm^3, $b_t = 0.08$ day^{-1}, $b_s = 0.04$ day^{-1}, $b_d = 0.04$ day^{-1}, $D = 1.25$ cm^2/day and $D_f = 0.393$ cm^2/day, $L = 0.0078$ cm.

References

1. Williamson K, McCarty PL (1976) A model of substrate utilization by bacterial films. 48(1):9–23
2. Sarkar S, Mazumder D (2015) Process design and application of aerobic hybrid bioreactor in the treatment of municipal wastewater. Int J Chem Nucl Mater Metall Eng World Acad Sci Eng Technol 9(3):466–470
3. Lee C-Y (1992) Model for biological reactors having suspended and attached growths. J Environ Eng 118(6):982–987
4. Gebara F (1999) Activated sludge biofilm waste water treatment system. Water Res 33(1):230–238

Chapter 5
Fabrication of Fixed-Bed Hybrid Bioreactor and Hydraulic Studies

5.1 Reactor Setup

5.1.1 Description of an Aerobic Hybrid Bioreactor Setup

For performance study of the fixed-bed hybrid bioreactor as well as for validation of the proposed mathematical model, initiative was taken up to fabricate two laboratory-scale fixed-bed hybrid bioreactors. One reactor was used in batch mode of operation and it consisted of a PVC bucket with aeration facility. Another reactor consisting of a PVC Jar with aeration facility and a secondary clarifier arrangement was used in continuous mode of operation.

5.1.1.1 Reactor Setup for Batch Study

A 10 L capacity PVC bucket was used as a laboratory-scale fixed film aerobic hybrid bioreactor as shown in Fig. 5.1. Perspex sheet of 3 mm thickness was provided as attachment surface for enhancing the growth of microorganism in biofilm. Nine equal sized fins made of perspex sheets and having height 275 mm were mounted centrally with a uniform angular slot of 40 degree. Total fixed and varying area of the attachment surface is measured to be 0.27 m^2 and 0.132 m^2, respectively. Two numbers of Aquarium pump were deployed for supplying air into the reactor.

5.1.1.2 Reactor Setup for Continuous Study

A 15-L capacity PVC jar was adopted as the aeration basin of the reactor. Perspex sheet (3 mm thickness) was taken as a raw material for fabrication of attachment surface for the growth of biomass. A frame of nine equal angular slots each of 40 degree and height of 275 mm made of perspex sheets were affixed together along the longitudinal sides as shown in Figs. 5.1 and 5.2.

S. Sarkar and D. Mazumder, *Fixed Bed Hybrid Bioreactor*, Green Energy and Technology,
https://doi.org/10.1007/978-981-33-4546-1_5

Fig. 5.1 Plan view of biofilm attachment surface

Fig. 5.2 Attached-growth setup without biofilm

The attachment frame was inserted in the jar with a spacing of 25 mm from bottom surface of the jar for accommodating the sludge growth at bottom. The projected arm of each finished fixed angular frame is of dimension 54 mm × 275 mm × 3 mm. Nine numbers rectangular fins of each size 52 mm × 275 mm × 3 mm were inserted inside the slots formed between two fixed angular fins with the projected arm of dimension 26 mm × 275 mm × 3 mm from the fixed surface. These rectangular fins acted as the variable attachment surface inside the fixed one. Total fixed surface area for attachment of biomass was 0.27 m^2 and total varying surface area for attachment was 0.132 m^2. A MS junction box is installed on the top of the jar through which nine (9) nos of 10 mm dia CPVC pipes were vertically placed through nine (9) slots of the biofilm surface, i.e., perspex sheets. A common 10 mm dia PVC pipe was fitted at top of the junction box through which the aeration was provided to the reactor prior to distribution through nine (9) nos of vertical pipes. One inlet CPVC pipe of 10 mm diameter and one outlet CPVC pipe of the same diameter were connected in the reactor in opposite direction to each other. One sludge outlet perpendicular to the direction of inlet–outlet line was at bottom of the reactor with a stopper as shown in Fig. 5.3.

One outlet pipe emerging from the reactor was extended to the secondary clarifier (PVC made, 5.5-L capacity) which was placed in a PVC made tub for collecting the overflow from the secondary clarifier. One outlet was also connected at the bottom of the tub for discharging the effluent from the system. The reactor was allowed to run by means of peristaltic pump as per HRT provided until quasi-steady-state condition reached. An air pump (maximum capacity 50 lpm) was used for providing aeration in the reactor. The schematic diagram of reactor setup is shown in Fig. 5.4.

Fig. 5.3 Assembly of hybrid bioreactor

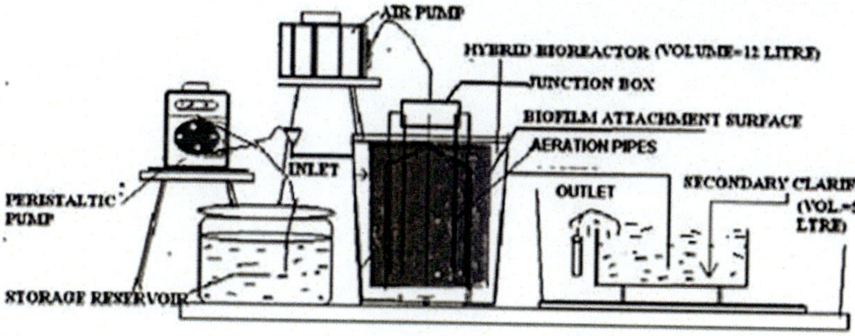

Fig. 5.4 Schematic diagram of a fixed film aerobic hybrid bioreactor under

5.2 Hydraulic Study on Fixed-Bed Hybrid Bioreactor

Tracer response study was performed to assess the hydraulic performance of the full-scale reactor used for the treatment of municipal wastewater. The objective of the study was to establish the hydraulic regime of the laboratory-scale fixed-bed hybrid bioreactor. The use of inert tracers for developing the residence time distribution curves is one of the simplest and well-established method for ascertaining the hydraulic regime of the full-scale reactor under any given operating condition.

5.2.1 *Approach*

The tracer study was conducted under slug input of the tracer at the inlet point and concurrent measurement of tracer concentration at the outlet point. For this purpose, lithium chloride (LiCl) as a tracer was used which is very much compatible with the biological environment. After initiation, at an interval of 15 and 30 min the effluent was collected from the secondary clarifier and the tracer concentration for the same was measured for a series of grab samples. The tracer study was conducted for the MLSS concentration in the range of 2000–2500 mg/l and HRT between 4 and 8 h. Tracer study in each run was continued until two successive tracer concentrations remain more or less same. In order to set any HRT say 4 h, the flow rate was adjusted by means of a peristaltic pump. The amount of tracer concentration was varied from 100 to 500 mg/l. Finally, tracer response curves were drawn in respect of C/C_0 versus t/t_o profile, where, C_0 and C are initial and final tracer concentration corresponding to t_0 and t, respectively, where t is the actual detention time and t_0 is the theoretical detention time.

5.2.2 *Theoretical Aspect of Tracer Response Curves*

For an ideal flow in completely mixed reactor, the ideal tracer response profile can be represented by [1]

$$C = C_0 e^{-(t/t_0)} = C_0 e^{-\theta}$$

Or,

$$C/C_0 = e^{-\theta} \tag{5.1}$$

C_0 = Initial concentration of the tracer in the reactor (mg/l).

C = Concentration of the tracer in the reactor at time t (mg/l).

θ = normalized detention time, i.e., t/τ, where, t = time (minutes), τ = theoretical detention time (minutes).

Now, theoretical tracer response curve (for completely mixed flow) was plotted for tracer concentration 100 mg/l, 400 mg/l and 500 mg/l with hydraulic detention time 4 h and MLSS concentration 2000 mg/l in Fig. 5.5 below.

The same curve is plotted for tracer concentration 100, 200 and 300 mg/l with detention time 6 h and MLSS concentration 2500 mg/l in Fig. 5.6.

Similarly, tracer response curves are also plotted for tracer concentration 100 and 200 mg/l under hydraulic retention time 8 h and MLSS concentration 2500 mg/l in as shown in Fig. 5.7.

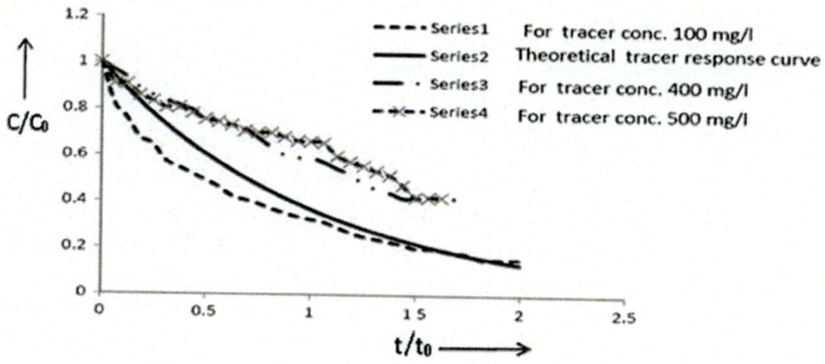

Fig. 5.5 Tracer response curves for different tracer concentrations at HRT = 4 h and theoretical tracer response curve

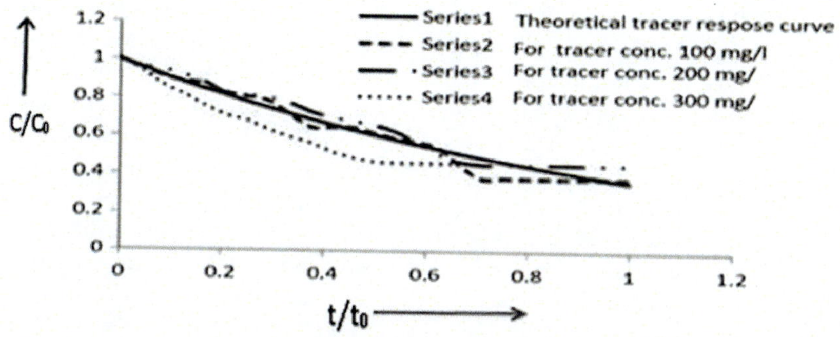

Fig. 5.6 Tracer response curves for different tracer concentrations at HRT = 6 h and theoretical tracer response curve

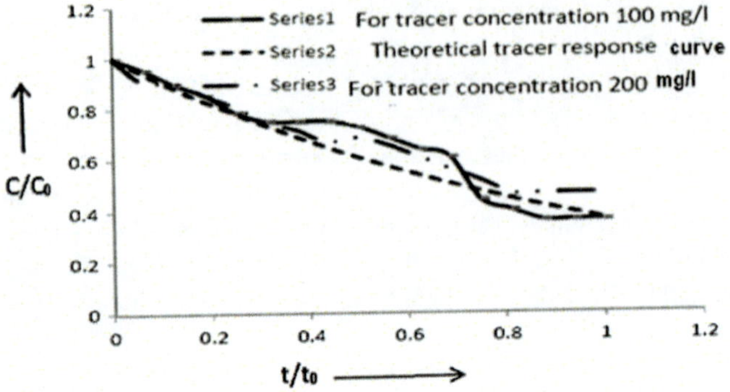

Fig. 5.7 Tracer response curves for different tracer concentrations at HRT = 8 h and theoretical tracer response curve

5.2.3 *Experimental Tracer Response Curves*

Tracer response study was performed at laboratory considering various combinations of tracer concentrations and hydraulic retention times as well as MLSS concentration.

.The output results of the different combinations are demonstrated in Table 6a of Annexure III-c. In every case, tracer response curves have been drawn considering t/t_0 in x-axis and C/C_0 in y-axis. Finally, all the experimental tracer response curves were superimposed with the theoretical ones. The superimposed curves for three cases of HRTs, like for 4, 6 and 8 h are shown in Figs. 5.5, 5.6 and 5.7.

One set of runs was conducted with tracer concentration of 100 mg/l under detention time of 4, 6 and 8 h in presence of MLSS concentration 2000 mg/l, 2500 mg/l and 2500 mg/l, respectively. The second set of runs was operated with tracer concentration of 200 mg/l under detention time of 6 and 8 h in presence of MLSS concentration 2500 mg/l. The third run was operated with tracer concentration of 300 mg/l under detention time of 6 h in presence of MLSS concentration 2500 mg/l. The fourth run was operated with tracer concentration of 400 mg/l under detention time of 4 h in presence of MLSS concentration 2000 mg/l. Lastly, the fifth run was operated with tracer concentration of 500 mg/l under detention time of 4 h in presence of MLSS concentration 2000 mg/l. All these combinations were chosen considering the most possible operating condition of the reactor in terms of MLSS and detention time.

5.3 Preliminary Studies for Comparison of Performance Between Fixed-Bed Hybrid Bioreactor and Suspended-Growth System

5.3.1 Basic Approach

As a preliminary study, two reactors—one fixed-bed hybrid bioreactor and another suspended-growth reactor were run with synthetic carbonaceous wastewater in the same operating conditions of input COD and MLSS at same batch period. The purpose of this study is to observe the performance of the fabricated hybrid bioreactor with respect to the suspended one. The batch period was provided as 8 h for all the cases. The sample was taken at a batch interval of 2 h. At each batch interval, MLSS of the reactor content, COD, $NH4^{+}$–N and $NO3^{-}$N of soluble sample were determined to compare the performance between the two reactors.

5.3.2 Experimental Procedure

Firstly, two reactors (two PVC Jars each of 15 L each) one suspended growth and other hybrid bioreactor were placed side by side connected with two numbers aqua pumps in each reactor. Required quantity of sludge from municipal wastewater was put in both the reactors. Synthetic carbonaceous wastewater with required input COD was prepared from the stock solution and accordingly was poured into each reactor. Six (6) batch runs were performed under synthetic carbonaceous wastewater with COD in the range of (100–250) mg/l and MLSS in the range of (2000–2500) mg/l. The batch study for both the suspended and hybrid bioreactor system was conducted for a total batch period of 8 h and was continued over 6 days. The results of the experimental results are shown in Table 6b, Annexure III-c. The results of preliminary batch study clearly demonstrate that the hybrid bioreactor system is more capable to remove especially the carbonaceous matter than the conventional ASP system. However, nitrification by means of oxidation of $NH4^{+}$–N has not been improved in case of fixed-bed hybrid bioreactor.

Reference

1. Metcalf and Eddy (2010) Wastewater Engineering, A text book, Reprint 21 st.

Chapter 6
Performance Evaluation of Fixed-Bed Hybrid Bioreactor Under Synthetic Carbonaceous Wastewater

6.1 Preparation of Synthetic Carbonaceous Wastewater

A synthetic carbonaceous wastewater was used as a stock solution, where, COD concentration was set to about 1000 mg/l. Adequate nitrogen and phosphorus were also added to this stock solution in accordance with carbon. Variable initial COD concentrations were set by diluting the stock solution with distilled water. The composition of stock synthetic solution is shown in Table 6.1.

6.2 Semi-batch Study in the Hybrid Bioreactor

Semi-batch study was conducted on the fixed-bed hybrid bioreactor using synthetic carbonaceous wastewater in order to get the response under highly biodegradable environment. This is also to ascertain the consistent performance of the fixed-bed hybrid bioreactor before going to its continuous operation. The objective, scope and detailed procedure of the semi-batch study are illustrated below.

6.2.1 Objective and Scope

The experimental plan was to check the response of the fixed-bed hybrid bioreactor under semi-batch mode of operation with synthetic wastewater. The semi-batch study was basically conducted as a part of performance evaluation of the fixed-bed hybrid bioreactor as well as for determining the kinetic coefficients from the experimental results required for finding out the outputs from the mathematical model.

Scope of the semi-batch study is to determine the COD concentration as well as mixed liquor suspended solid (MLSS) in the respective batch intervals during the period of semi-batch operation. Apart from that, the attached biomass density

S. Sarkar and D. Mazumder, *Fixed Bed Hybrid Bioreactor*, Green Energy and Technology,
https://doi.org/10.1007/978-981-33-4546-1_6

Table 6.1 Composition of synthetic carbonaceous wastewater with COD concentration 1000 mg/l (approx.)

Compound	Concentration (mg/l)
Dextrose ($C_6H_{12}O_6$)	940
NH_4NO_3	260
KH_2P0_4	82.67
$CaCl_2$	27.5
$MgSO_4.7H_20$	22.5
$FeCl_3$	0.25

was also measured time to time in this semi-batch study. Finally, semi-batch study gives the optimum operating conditions like optimum HRT and initial MLSS values, which are required for running the continuous operation of the reactor in the selected mode. Ammonium and nitrate-nitrogen of the influent and effluent sample were also measured to check any occurrence of nitrification/denitrification in the reactor.

6.2.2 *Procedure of Semi-batch Study*

Semi-batch study was conducted in the laboratory-scale fixed-bed hybrid reactor at an interval of 2 h for the batch period of maximum 8 h. The study was conducted for total 7(seven) days. The influent substrate concentration was varied from 100 to 250 mg/l with an increment of 50 mg/l at each time in accordance with the tentative COD concentration in the municipal wastewater. Different biomass concentration (MLSS) was set out at the initial stage of batch study, which varied from 2000 to 2500 mg/l. During this semi-batch study, at each batch interval effluent sample was collected and the relevant parameters including MLSS and COD (substrate) were measured. In addition, biomass sample was extracted from the attached surface to measure the density of the attached biomass. Biomass density in the attached surface was determined by washing the biofilm surface with 25 ml 0.1 (N) NaOH and thereafter heating the solution at 80 degree Celsius [1] followed by spectrometric observation as per Lowry's method. The total biomass in the biofilm surface bioreactor ($X_{attached}$) was determined from the principle of Lowry's method. Apart from COD, effluent soluble sample was also analyzed for finding out the ammonium-nitrogen concentration (NH_4^+–N) and nitrate-nitrogen concentration (NO_3^-N) with the help of ion meter. The purpose of the analysis was to check the occurrence of nitrification and denitrification in the fixed-bed hybrid reactor. The results of semi-batch study with synthetic wastewater are shown in Table 6.2.

Table 6.2 Results of 8-days semi-batch study conducted in batch interval @ 2 h

Day	Influent	Initial	Batch	MLSS(mg/l)	COD(mg/l)	NH_4^+–N	NO_3–N	Attached
1	100	2000	0	1958	95	6.32	20	878
			2	1970	75	3.1	23	1145
			4	1978	60	1.3	23.5	1398
			6	1984	48	0.9	24.7	1475
			8	2000	25			1579
2	100	2500	0	2533	97	4.8	20.4	885
			2	2541	87	2.9	22.3	1475
			4	2556	67	2.6	22.5	1650
			6	2570	45	2.3	23	1883
			8	2579	29			1917
3	150	2000	0	1989	165	16.5	23	672
			2	2015	130	13	25.5	1588
			4	2041	86	10.6	26	1889
			6	2050	46	8.3	31	1975
			8	2068	19	6.65	33.5	2010
4	150	2500	0	2490	158	12	28	680
			2	2531	90	7.5	34	1818
			4	2541	45	4.5	37	2151
			6	2545	27	3.0	38	2194
			8	2547	18	5.0	15	2244
5	200	2500	0	2575	189	20	28	979
			2	2625	87	15	34	1971
			4	2630	41	11.5	37	2023
			6	2595	30	11.5	38.5	2152
			8	2614	13	10.2	41	2509
6	250	2000	0	2067	270	24	42	999
			2	2118	155	14.2	48	1589
			4	2122	59	9.3	51	1693
			6	2093	28	6.5	54.3	1834
			8	2116	21	3.6	53	1990
7	250	2500	0	2455	245	18.5	43	1050
			2	2528	139	13.5	47	1985
			4	2561	89	11.5	50	2065
			6	2578	61	8.3	50	2196
			8	2590	38	7.95	52.5	2575

6.3 Details of Continuous Study

The performance of hybrid bioreactor was also evaluated under continuous mode of operation using the same synthetic carbonaceous wastewater as used in the semi-batch study. In order to accomplish this, flow rate was adjusted with the help of peristaltic pump complying with desired hydraulic retention time (HRT). Apart from HRT, other operating parameters were also varied accordingly to match with the most possible operating regime for the municipal wastewater. Analysis of the influent and effluent samples for soluble COD (mg/l) expressed the performance of hybrid bioreactor.

6.3.1 Basic Approach

A continuous study was planned with synthetic carbonaceous wastewater as influent with varying HRT ranging between 4 and 8 h, variable influent organic strength (COD) between 150 and 250 mg/l and varying input biomass concentrations in the range of 20,003,500 mg/l. There are 27 (twenty-seven) numbers of continuous runs, performed on the reactor with 3 (three) variations in each of influent COD concentrations, HRT and MLSS concentrations The reactor was run under continuous mode, following the above-stated operating ranges with a view to find out the COD removal efficiency as a performance indicator. The continuous study was also aimed to validate the present mathematical model as developed earlier. Each continuous run under a particular operating condition was monitored until the quasi-steady-state condition reached in the reactor system.

6.3.2 Procedure for Continuous Study

The hybrid bioreactor was run under continuous mode with three variable initial COD concentrations, viz 150, 200 and 250 mg/l and maintaining a HRT of 4, 6 and 8 h. The initial biomass concentration (both suspended and attached) was also varied in commensurate with initial COD concentration. The reactor was allowed to run by means of peristaltic pump until quasi-steady-state condition reached. The influent and effluent sample was taken for measuring final COD concentration, whereas the reactor content was taken for determining steady-state suspended biomass concentration. A portion of attached media (perspex sheet) was also washed out with 0.1 N NaOH solution to measure the attached biomass.

All the relevant parameters except MLSS were measured as per Standard Methods (1995). In case of MLSS, 50 mL mixed liquor sample was taken out from the reactor and the same was filtered. The filtrate retained in the filter paper was allowed to dry in a hot air oven at 105 °C until it attained a constant weight. Then the filter paper was put into a desiccator and its final weight was taken. Thus, it is possible to calculate the MLSS by subtracting the initial weight from the final one.

Biomass density in the attached media was determined by washing the biofilm surface with 25 mL 0.1 (N) NaOH and then heating the solution at 80 °C [1, 2]. Subsequently, by measuring the protein concentration in the washed out content as per Lowry's method [3] the amount of attached biomass was estimated.

The results of continuous study with synthetic wastewater are tabulated in Table 6.3.

Table 6.3 Continuous study with synthetic wastewater sample

Run No.	HR	MLSS	INITIAL	F/M	Effluent	% age	Loading
01	4	2500	150	0.1464	89	41	0.9
02		3000		0.15	75	50	0.9
03		3500		0.163	55	63	0.9
04	6	2500		0.168	42	70	0.6
05		3000		0.147	40	73	0.6
06		3500		0.128	38	75	0.6
07	8	2500		0.15	25	83	0.45
08		3000		0.13	20	87	0.45
09		3500		0.114	17	89	0.45
10	4	2500	200	0.26	92	54	1.2
11		3000		0.22	89	56	1.2
12		3500		0.238	61	70	1.2
13	6	2500		0.24	43	75	0.8
14		3000		0.207	45	78	0.8
15		3500		0.184	39	81	0.8
16	8	2500		0.2076	27	87	0.6
17		3000		0.179	21	90	0.6
18		3500		0.156	18	91	0.6
19	4	2500	250	0.36	100	60	1.5
20		3000		0.324	88	65	1.5
21		3500		0.31	69	72	1.5
22	6	2500		0.312	38	78	1
23		3000		0.268	49	80	1
24		3500		0.234	45	82	1
25	8	2500		0.264	28	88	0.75
26		3000		0.225	25	90	0.75
27		3500		0.197	20	92	0.75

6.4 Compatibility of Fixed-Bed Hybrid Bioreactor for the Treatment of Synthetic Carbonaceous Wastewater Treatment

The present continuous study has been carried out to evaluate the performance of a fixed-bed aerobic hybrid bioreactor for removal of carbonaceous organic matter from synthetic carbonaceous wastewater. Various experimental combinations have been performed to obtain optimum results in terms of COD removal. COD removal profiles obtained from various experimental studies suggest that more than 85% COD removal is possible with initial COD concentration in the range of (150 ± 50) to (250 ± 50) mg/l in fixed-bed hybrid bioreactor at a hydraulic retention time of 8 h. Optimum results have been obtained at MLSS concentration (3000–3500) mg/l and at a hydraulic retention period (6–8) hours corresponding to attached biomass concentration of about 1000 mg/l. The rate of biodegradation may be enhanced by means of additional biomass present in the attached-growth surface, in turn by increasing the total biomass concentration.

References

1. Lazarova V, Pierzo V, Fontvielle D, Manem J (1994) Integrated approach for biofilm characterization and biomass activity control. Water Sci Technol 29(7):345–354
2. Cox HJ, et al (1999) Chemical removal of biomass from waste air biotrickling filters: Screening of chemicals of potential interest. Water Res 33:2383–2391
3. Lowry OH, Rosebrough NJ, Fair AL, Randall RJ (1951) Protein measurement with Folin Phenol Reagent. J Biol Chem 193:265–275

Chapter 7
Treatability Study of Municipal Wastewater in Fixed-Bed Hybrid Bioreactor

7.1 Criteria for Selection of Wastewater

In the medium and small town, the strength of wastewater varies mostly due to disintegrated sewerage systems. In case of low to medium strength municipal wastewater, main problem for treatment is sustaining the biomass in the reactor without the return sludge, i.e., settled biomass from the secondary clarifier. Presently, adequate biomass is maintained through extended aeration and with sludge recirculation, which increases the overall cost of treatment on account of pumping. In case of especially low strength municipal wastewater, which is highly biodegradable, the suspended biomass in the reactor tends to get washed out rapidly if there is no recirculation of settled biomass. Thus, the introduction of attached biomass into the suspended-growth system can be reasonably thought for maintaining sufficient biomass in the reactor. The footprint area of such integrated system, i.e., hybrid bioreactor is thus small compared to the conventional ASP, thereby making the economy of the treatment system.

7.2 Modality of Treatment of Municipal Wastewater in the Laboratory-Scale Reactor

For performance evaluation of the reactor as well as for validation of the proposed mathematical model, initiative was taken up for conducting both the batch and continuous mode operation of the fixed-bed hybrid bioreactor. The laboratory-scale hybrid bioreactor was run under semi-batch mode using municipal wastewater for determining the kinetic coefficients based on the graphical method as developed from the mathematical modeling. Continuous study was performed with the municipal wastewater under various hydraulic retention time (HRT) and biomass concentration. The validation of the mathematical modeling with the real municipal wastewater was done considering the kinetic coefficients derived from the semi-batch study.

S. Sarkar and D. Mazumder, *Fixed Bed Hybrid Bioreactor*, Green Energy and Technology, https://doi.org/10.1007/978-981-33-4546-1_7

Fig. 7.1 Process of collection of municipal wastewater from the Pit in pumping station

7.3 Description of Collection Point

Domestic wastewater from municipal wards of Kolkata Municipal Corporation (KMC) is conveyed through a sewerage system and collected in various pumping stations. In the present study, raw municipal wastewater was collected from a pumping station of Kolkata Municipal Corporation located at Sealdah, West Bengal, India. The wastewater was collected from underground pit, where the wastewater from different places under respective sewerage zones is allowed to mix together. The grab sample was collected from the pit manually to PVC jars as shown in Fig. 7.1.

7.4 Modality of Collection and Preservation Steps

Initially, municipal wastewater was collected periodically @ 30 L per day (carrying each jar of 10 L) in two phases, Phase A and Phase B for characterization of raw wastewater and biological acclimation with raw wastewater, respectively. Phase A was continued from the summer period and Phase B was continued from the monsoon period.

For the preservation of wastewater during characterization, biomass acclimation and batch as well as continuous study, 1(one) ml of conc. H_2SO4 was added per liter of wastewater until pH value dropped below 2.0 and the same was kept in the refrigerator. At the time of operation of the hybrid bioreactor with preservative wastewater, necessary adjustment of pH up to about 7.2 was made with the help of filtered lime solution.

7.5 Characterization of Municipal Wastewater

Before going for any laboratory study on municipal wastewater, the same was characterized for all the relevant parameters. The parameters selected for characterization of the raw municipal wastewater are pH, total alkalinity, chemical oxygen demand (COD), biochemical oxygen demand (BOD3 at 27 °C), total solids (TS), total suspended solids (TSS), total dissolved solids(TDS), total Kjeldahl Nitrogen (TKN), PO_4^{3-} and NO_3^{2-}. The purpose of characterization of wastewater was to study the nature of the same and to know the present strength of the wastewater. The detailed characterization of the municipal wastewater was performed in the laboratory following the Standard Methods (1998) and the results of same are tabulated in Table 7.1.

7.6 Acclimation of Biomass and Its Enrichment

For the sake of acclimation, the sludge from one ASP unit was collected and then it was fed with gradually increasing amount of municipal wastewater. The municipal wastewater was fed under batch mode to a bucket of 10-L liquid volume over 21 days. The progress of acclimation was monitored by taking sample at the beginning and at the end of a particular batch period. There are two phases of monitoring Phase A during summer and Phase B during monsoon.

The parameters, viz. pH, MLSS, COD and BOD3 at 27 °C were regularly measured at every 24 h interval from initial feeding. The purpose of monitoring in two phases is to see the effect of seasonal fluctuations on biological acclimation. It was observed in Phase B, i.e., in rainy season, the municipal wastewater was found diluted resulting in lower value of initial COD compared to Phase A. Under Phase A, initial COD value was in the range of 127–278 mg/l, whereas the final COD value ranged from 51–150 mg/l. The initial COD value for Phase B was observed to be between 64 and 112 mg/l, whereas the final COD value ranged from 12 to 26 mg/l. The BOD to COD ratio was found to be in the range of 0.7–0.8 during the acclimation study. The pH value was observed satisfactorily in the range between 7.1 and 7.7 in both Phase A and Phase B throughout the acclimation period. The MLSS (biomass) concentration was measured at the initial and the final stage of the batch acclimation and the average value was noted. The results of the acclimation study are shown in Table 7.2.

The variation of initial and final COD profile in the periodic batch study for Phase A is shown in Fig. 7.2 and that of Phase B is shown in Fig. 7.3.

Table. 7.1 Characterization of municipal wastewater collected from Palmer Buzzer, Kolkata Municipal Corporation

Date of collection	pH	Total alkalinity (mg/l)	COD (mg/l)	BOD (mg/l)	TS (mg/l)	TDS (mg/l)	TKN (mg/l)	PO_4^{3-} (mg/l)	NO_3^{2-} (mg/l)	TSS (mg/l)
Collection point: pumping station, Palmer Buzzer, Under KMC										
11.01.14	7.21	360	264	208	514	468	22.4	5.1	0.8	46
18.01.14	7.14	370	256	200	544	520	24	8.5	0.92	24
01.02.14	7.17	380	256	235	406	380	22.4	4.73	2.2	26
08.02.14	7.14	350	244	220	416	380	28	3.2	2.15	36
15.02.14	7.18	350	224	180	512	444	25.2	3.92	3.44	68
22.02.14	7.25	350	283	240	488	420	19.6	7.25	3.5	68
Statistical analysis										
Mean	7.18	360	255	214	480	435	23.6	5.45	2.43	44.67
Standard deviation	0.016	4.7	7.36	8.4	21	20.2	1.07	0.756	0.31	7.34
Max	7.25	380	283	240	544	520	25.2	8.5	3.44	68
Min	7.14	350	224	180	406	380	19.6	3.2	0.92	24

Table. 7.2 Results of monitoring during acclimation

Phase	Batch run no	Volume of municipal wastewater feed (l)	MLSS (mg/L)	Batch period (h)	pH	COD (mg/l)	BOD (mg/l)
Phase A	01	1	2665	0	7.5	149	119
				24		99	74
	02	1	2710	0	7.4	178	125
				24		128	99
	03	1	2700	0	7.5	162	122
				24		112	87
	04	1	2750	0	7.2	127	98
				24		77	62
	05	1	2775	0	7.7	152	116
				24		102	81
	06	1	2802	0	7.7	130	104
				24		80	64
	07	1	2988	0	7.6	170	129
				24		98	69
	08	1	2852	0	7.35	278	206
				24		150	117
	09	1	2774	0	7.5	160	123
				24		64	51
	10	1	2688	0	7.3	230	184
				24		121	93
	11	1	3138	0	7.6	160	128
				24		85	65
	12	1	3578	0	7.1	205	144
				24		51	36
	13	1	3900	0	7.35	200	152
				24		75	53
	14	1	3800	0	7.7	195	156
				24		65	46
	15	1	3300	0	7.7	176	132
				24		112	85
	16	1	4272	0	7.2	220	156
				24		67	48
Phase B	17	5	4300	0	7.3	112	85
				24		25	20

(continued)

Table. 7.2 (continued)

Phase	Batch run no	Volume of municipal wastewater feed (l)	MLSS (mg/L)	Batch period (h)	pH	COD (mg/l)	BOD (mg/l)
	18	5	4400	0	7.1	92	69
				24		21	15
	19	5	4575	0	7.7	88	62
				24		14	10
	20	5	4230	0	7.7	93	70
				24		26	21
	21	5	4335	0	7.2	64	51
				24		12	10

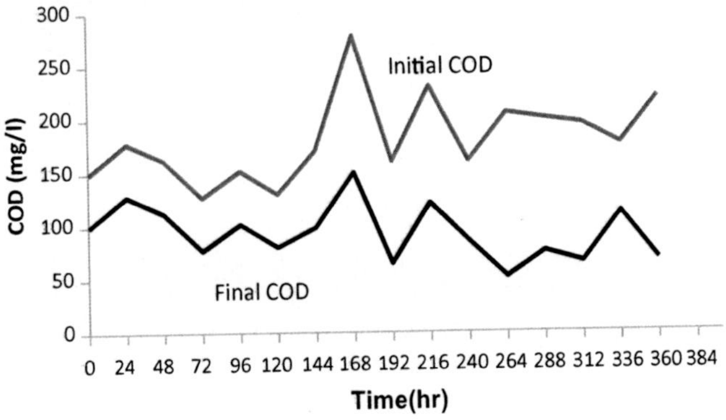

Fig. 7.2 Variation of initial and final COD profile in the periodic batch study in Phase A

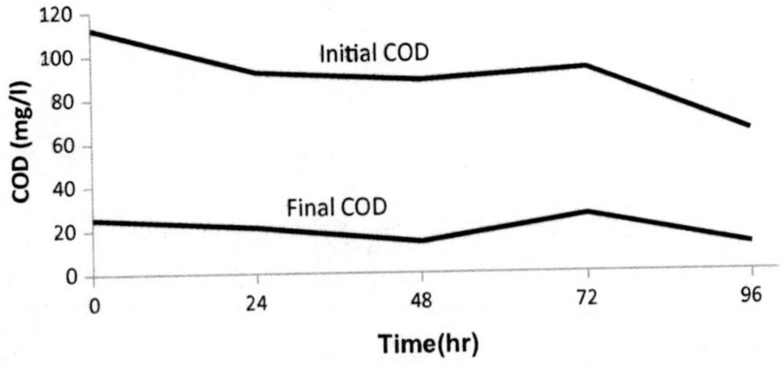

Fig. 7.3 Variation of initial and final COD profile in the periodic batch study in Phase B

7.7 Description of Semi-batch Study with Municipal Wastewater

The semi-batch study was conducted in the fixed-bed hybrid bioreactor, already fabricated using the municipal wastewater. For this purpose, the reactor was fed with varying strength of municipal wastewater under different initial MLSS concentrations. The performance of the reactor was monitored at certain intervals over a specific batch period.

7.7.1 Objective and Scope

The experimental plan was to check the response of the fixed-bed hybrid bioreactor under semi-batch mode of operation. The semi-batch study was basically conducted with an objective for performance evaluation of the fixed-bed hybrid bioreactor as well as for determining the kinetic coefficients from the experimental results, which are required for finding out the outputs from the mathematical model.

Scope of the semi-batch study was to determine the COD as well as mixed liquor suspended solid (MLSS) concentration at various batch intervals over a set of batch runs. At the same time, the attached biomass density was also measured at respective batch intervals.

7.7.2 Procedure of Semi-batch Study

Semi-batch study was conducted in the laboratory-scale hybrid fixed biofilm reactor at an interval of 3 h over a total 9 h batch period. Different biomass concentration (MLSS) was set out at the initial time of batch operation. During this study, the effluent sample was collected at each batch interval and analyzed for MLSS as well as COD (substrate) concentration. In addition, biomass sample was extracted from the attached surface to measure the density of the attached microorganisms. Biomass density in the attached surface can be determined by washing the biofilm surface with 25 ml 0.1 (N) NaOH and thereafter heating the solution at 80 °C [1] followed by spectrometric observation of protein as per Lowry's method. The total biomass in the biofilm bioreactor (Xattached) was determined from a calibration curve correlating protein and attached biomass. The results of semi-batch study with municipal wastewater are shown in Table 7.3.

Table. 7.3 Results of three days semi-batch study conducted in batch interval @ 3 h with municipal wastewater

Day	Batch period (h)	MLSS (X) (mg/l)	Attached biomass conc. (mg/l)	COD (mg/l)
1	0	1989	580	112
	3	2028	1368	55
	6	2037	2321	16
	9	2040	2332	5
2	0	2900	979	100
	3	2921	1384	50
	6	2940	2086	16
	9	2950	2216	5
3	0	4488	849	130
	3	4506	1792	70
	6	4523	1805	28
	9	4538	1832	15

7.8 Description of Continuous Study

The fixed-bed hybrid bioreactor was run under continuous mode for the treatment of municipal wastewater. For this purpose, the reactor was fed with municipal wastewater under varying hydraulic retention times (HRT) and MLSS. A peristaltic pump was employed to adjust the flow rate and thereby fixing a certain HRT.

7.8.1 Basic Approach

A continuous study was performed using municipal wastewater as influent with varying HRT ranging between 4 and 8 h and varying input biomass concentrations in the range of 2500–3000 mg/l. The reactor was run under continuous mode following the above-stated operating ranges with a view to find out the COD removal efficiency as a performance indicator. The continuous study was also aimed to compare the performance data with predicted ones and hence to validate the mathematical modeling developed earlier. Each continuous run under a particular operating condition was monitored until the quasi-steady-state condition reached in the reactor system.

7.8.2 Procedure for Continuous Study

There are 9 (nine) numbers of continuous runs each of 72 h performed on the reactor with 3 (three) variations of HRT and three (3) variations of input biomass concentrations. The reactor was allowed to run by means of peristaltic pump until quasi-steady-state condition reached. The adequate aeration was ensured in the reactor with the high-flow air pump. The effluent sample was taken from the overflow of the secondary clarifier for measuring final COD concentration, whereas the reactor content was taken for determining steady-state suspended biomass concentration. A portion of attached media (perspex sheet) was also washed out with 0.1 N NaOH solution to measure the protein concentration and thereby the attached biomass. The results of continuous study with municipal wastewater are shown in Table 7.4.

7.9 Compatibility of Fixed-Bed Hybrid Bioreactor for Municipal Wastewater Treatment

The present continuous study has been carried out to evaluate the performance of a fixed-bed aerobic hybrid bioreactor for removal of carbonaceous organic matter from municipal wastewater. Various experimental combinations have been made to obtain optimum results in terms of COD removal efficiency. COD removal profiles obtained from various experimental studies suggest that more than 85% COD removal is possible with initial COD concentration 151 mg/l and initial MLSS concentration 3000 mg/l at a hydraulic retention time of 8 h. Thus, optimum results are obtained at MLSS concentration 3000 mg/l and at a hydraulic retention period 8 h corresponding to attached biomass concentration of 1000 mg/l. The rate of biodegradation may be enhanced by means of additional biomass present in the attached surface, in turn by increasing the total biomass concentration.

Table. 7.4 Results of continuous study with municipal wastewater

Run No	HRT (h)	MLSS (mg/l)	INITIAL COD (mg/l)	EFFLUENT COD Sw(mg/l) from	F/MRATIOQ (S0-Sw)/VX(kg/kg/day)	Loading rate (Kg/cum/day)*Q* * S0/V	%age removal efficiency
01	4	2000	120	47.5	0.2175	0.72	60.4
02		2500	134	49	0.204	0.604	63.4
03		3000	145	52.2	0.1856	0.87	64
04	6	2000	118	18.8	0.1984	0.472	84
05		2500	134	22	0.1792	0.536	83.6
06		3000	160	23.5	0.182	0.64	85.3
07	8	2000	110	11.7	0.14745	0.33	89.4
08		2500	147	12	0.162	0.44	91.83
09		3000	151	12.3	0.1387	0.453	91.85

Reference

1. Lazarova V, Pierzo V, Fontvielle D, Manem J (1994) Integrated approach for biofilm characterization and biomass activity control. Water Sci Technol 29(7):345–354

Chapter 8
Experimental Validation of the Model Developed and Process Design of Fixed-Bed Hybrid Bioreactor

8.1 Basic Consideration

The FORTRON program developed for mathematical modeling was run considering the input values like influent substrate concentration ($S0$), hydraulic retention time (θ), specific surface area (a), attached biomass density (Xf) and relevant kinetic coefficients as well as biofilm specific parameters like half-velocity constant (K), maximum specific rate of substrate utilization (k), biomass yield coefficient (Y), biomass loss rate due to shear from biofilm (b_s), biomass decay rate (b_d), total biomass loss rate (b_t), molecular diffusivity in biofilm (Df) and thickness of liquid layer in biofilm (L). The program was run for predicting the exiting substrate (COD) concentration, which is the key output parameter for performance evaluation of the reactor. In this case, experimental value of the effluent COD concentrations flux and total biofilm thickness was experimentally found out in semi-batch study and was compared with that obtained from computer program output for validation of the proposed under different conditions was validated with that predicted from the bioreactor model. In order to find out the accuracy of the developed model, one laboratory-scale hybrid bioreactor setup was run both under semi-batch and continuous mode. The continuous mode run was conducted using both synthetic carbonaceous wastewater and real municipal wastewater. The output data like exit substrate concentration, average substrate flux and total bio-film thickness were experimentally found out in semi batch study and was compared with that obtained from computer program output for validation of the proposed model. However, only exiting substrate concentration has been considered for validation under continuous mode of operation using both synthetic carbonaceous wastewater and real municipal wastewater.

S. Sarkar and D. Mazumder, *Fixed Bed Hybrid Bioreactor*, Green Energy and Technology, https://doi.org/10.1007/978-981-33-4546-1_8

8.2 Methodology for Experimental Validation

Firstly, the kinetic coefficients required as inputs for running the mathematical model were determined experimentally at laboratory. The technical process for the determination of kinetic coefficients from the semi-batch study is demonstrated below. After finding out the kinetic coefficients from semi-batch study, the mathematical model for hybrid bioreactor was validated with the data available from continuous study for both synthetic carbonaceous and real municipal wastewater. Thereafter, the experimental effluent COD concentrations have been plotted with respect to those predicted from the model to examine the accuracy of the developed model.

8.2.1 Determination of Kinetic Coefficients for Synthetic Carbonaceous Wastewater Treatment

8.2.1.1 Determination of K_s and μ_{max} (Ref. Sect. 4.2, Chapter 4)

Specific growth rate μ is calculated from the expression

$$\mu = \left(X_{\text{hybrid 2}} - X_{\text{hybrid 1}}\right)\left(X_{\text{hybrid 1}}\right)\theta$$

as shown in Table 8.1.

In order to determine the half-saturation constant (K_s) and maximum specific growth rate (μ_{max}) for synthetic carbonaceous wastewater treatment in the hybrid bioreactor system, μ was plotted along Y-axis against S_w plotted in X-axis as shown in Fig. 8.1 of Annexure VI-a. A vertical intercept equals $\mu_{max}/2$, is drawn intersecting the "μ" versus "S_w" curve at a point, wherefrom a perpendicular can be drawn to get the value of "K_s" as shown in Fig. 8.1.

It is evident from Fig. 8.1 of Annexure VI-a, $\mu_{max} = 0.35\ h^{-1}$=8.4 day^{-1}, K_s=40 mg/l=0.040 mg/cc.

8.2.1.2 Determination of K and Y

In order to determine the k, substrate mass balance in hybrid bioreactor at steady-state condition was considered. As a result, from the model equation of substrate mass balance, different values of $X\theta S_w/\{(S_0 - S_w)(K_s + S_w)\}$ were plotted against $\theta/(S_0 - S_w)$ corresponding to different hydraulic retention times, and different MLSS and COD concentrations (Ref. Sect. 3.2, Chapter 3).

Now, the values of $\theta/(S_0 - S_w)$ and $X\theta S_w/\{(S_0 - S_w)(K_s + S_w)\}$ are shown in Fig.8.2.

Accordingly, a best-fit line was drawn for this scatter diagram as shown in Fig. 8.2 with a relationship $y = -0.046x + 0.086$. Considering $p = 0.93$, the value of k was

Table 8.1 Determination of specific growth rate of biomass "μ" in a batch study (for synthetic carbonaceous wastewater)

	Batch period (θ, *h*)	Biomass	COD (S_W, mg/l)	$\mu(h^{-1})$
At first day of batch study				
9 a.m.	0	2526	158	
11 a.m.	2	4131	90	0.318
1 p.m.	4	4434	45	0.18877791
3 p.m.	6	4476	27	0.12861227
5 p.m.	8	4522	18	0.09873529
At second day of batch study				
9 a.m.	0	2627	189	
11 a.m.	2	4534	87	0.32872600
1 p.m.	4	4410	41	0.16966900
3 p.m.	6	4489	30	0.11809386
5 p.m.	8	4822	13	0.10442179
At third day of batch study				
9 a.m.	0	2120	270	
11 a.m.	2	3516	155	0.32921123
1 p.m.	4	3612	59	0.17586828
3 p.m.	6	3707	28	0.12471937
5 p.m.	8	3867	21	0.10298875

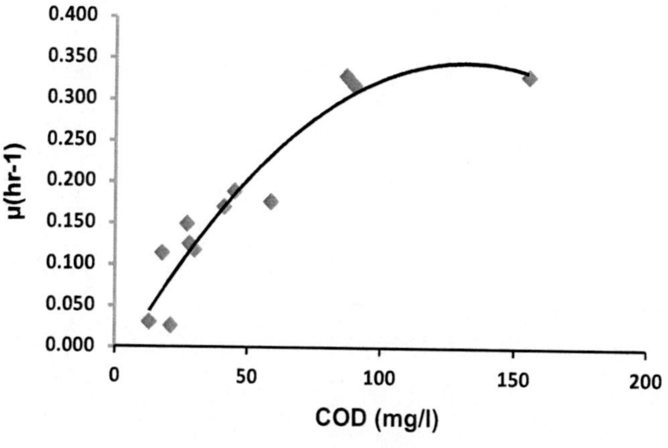

Fig. 8.1 Biomass growth curve for the determination of half-velocity constant *K* (for batch study with synthetic carbonaceous wastewater)

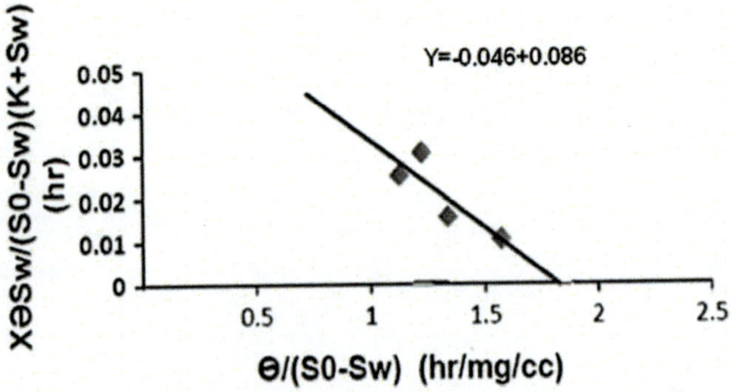

Fig. 8.2 Steady-state substrate balance for hybrid growth for the determination of k and Y (for batch study with synthetic carbonaceous wastewater)

determined from the intercept value. In addition, the yield coefficient, i.e., Y, was obtained from the relationship $Y = \mu_{max}/k$.

Now, from Fig. 8.2 1/pk = 0.086 day, p = porosity = 0.93. Hence, $k = 12.5\ \text{day}^{-1}$, $Y = \mu_{max}/k = 8.4/12.5 = 0.7$.

The value of Y showed a higher value in comparison with that for municipal wastewater, as stated in earlier literatures, presumably exhibiting high rate of organic conversion.

8.2.1.3 Determination of B_s, B_t and B_d

In order to determine b_s, b_t and b_d, the biomass balance for hybrid bioreactor system at steady-state condition was considered. It was represented by the equation YAJ/PX=$1/\theta_c - (YkS_w)/(K_s+S_w))$ (Ref. Clause 3.2, Chapter 3), where J = substrate flux within the attached-growth system, θ_c = solid retention time (SRT).

Now, θ_c is calculated as $p^*\theta$ for the case of no recirculation. The values of $(1/\theta_c - (YkS_w)/(K_s + S_w))$ and $(YaJ)/(pX)$ are presented in Fig. 8.3.

Accordingly, a best-fit line was drawn as shown in Fig. 8.3. The slope of the best-fit equation, i.e., $y = 15.08x + 2.18$, represents the values of $(b_s + b_d)/b_s$. On the other hand, the intersection of Y-axis represents the value $b_d(b_s+b_d)/b_s$. The coefficient of total biomass loss, i.e., b_t, can be calculated as $b_t = b_s + b_d$. From Fig. 8.3, m = 15.08 and c = 2.18, which shows biomass decay loss is 0.14 day^{-1}, biomass shear loss is 0.01 day^{-1} and total biomass loss is 0.15 day^{-1} (Ref. Sect. 3.2, Chapter 3).

Therefore, the kinetic coefficients and physical data used for prediction of the effluent substrate (COD) concentration under continuous study with synthetic wastewater are as follows:

$k = 12.5\ \text{day}^{-1}$, $Y = 0.7$, $K = 0.04\ \text{mg/cm}^3$, $b_t = 0.15\ \text{day}^{-1}$, $b_s = 0.01\ \text{day}^{-1}$, $b_d = 0.14\ \text{day}^{-1}$, $D_f = 0.64\ \text{cm}^2/\text{day}$ (assumed).

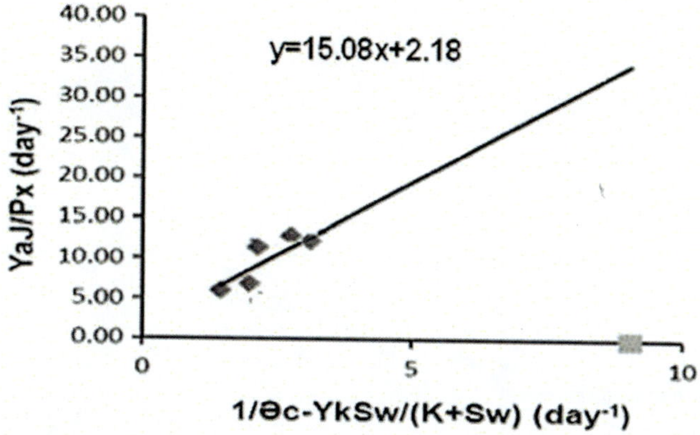

Fig. 8.3 Biomass balance for suspended growth for the determination of b_s, b_t and b_d (for batch study with synthetic carbonaceous wastewater)

8.2.2 *Determination of Kinetic Coefficients from Semi-batch Study with Municipal Wastewater*

8.2.2.1 Determination of K_s and μ_{max} (Ref. Clause 3.2, Chapter 3)

Specific growth rate μ is calculated from the expression

$$\mu = (X_{\text{hybrid2}} - X_{\text{hybrid1}})(X_{\text{hybrid1}})\theta$$

as shown in Fig. 8.4.

In order to determine the half-saturation constant (K_s) and maximum specific growth rate (μ_{max}) for synthetic carbonaceous wastewater treatment in the hybrid bioreactor system, μ was plotted along Y-axis against S_w plotted in X-axis as shown

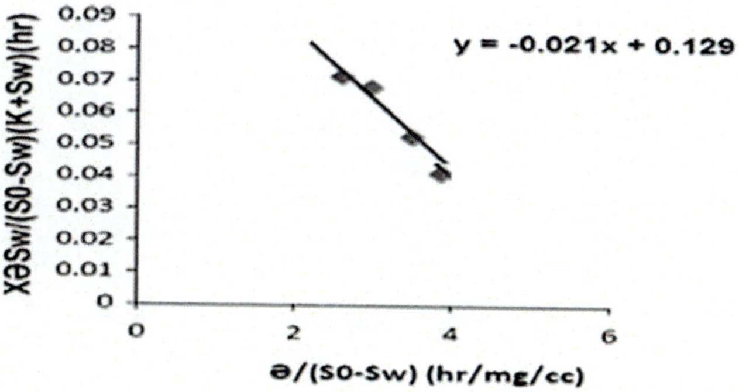

Fig. 8.4 Steady-state substrate balance for hybrid growth for the determination of k and Y (for batch study with municipal wastewater)

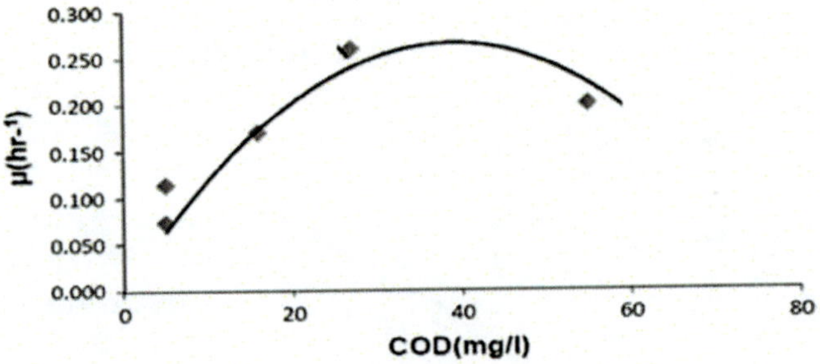

Fig. 8.5 Biomass growth curve for the determination of half-velocity constant K (for batch study with municipal wastewater)

in Fig. 8.5 of Annexure VI-a. A vertical intercept equals $\mu_{max}/2$, is drawn intersecting the "μ" versus "S_w" curve at a point, wherefrom a perpendicular can be drawn to get the value of "K_s" as shown in Fig. 8.5.

It is evident from Fig. 8.5, μ_{max}=0.265 h^{-1}=6.36 day^{-1}, K_s = 12 mg/l = 0.012 mg/cc.

8.2.2.2 Determination of K and Y

In order to determine the k, substrate mass balance in hybrid bioreactor at steady-state condition was considered. As a result, from the model equation of substrate mass balance, different values of $X\theta S_w/\{(S_0\text{-}S_w)(K_s + S_w)\}$ were plotted against $\theta/(S_0 - S_w)$ corresponding to different hydraulic retention times, different MLSS and COD concentrations (Ref. Sect. 3.2, Chapter 3).

Now, the values of $\theta/(S_0 - S_w)$ and $X\theta S_w/\{(S_0-S_w)(K_s + S_w)\}$ are shown in Fig. 8.6.

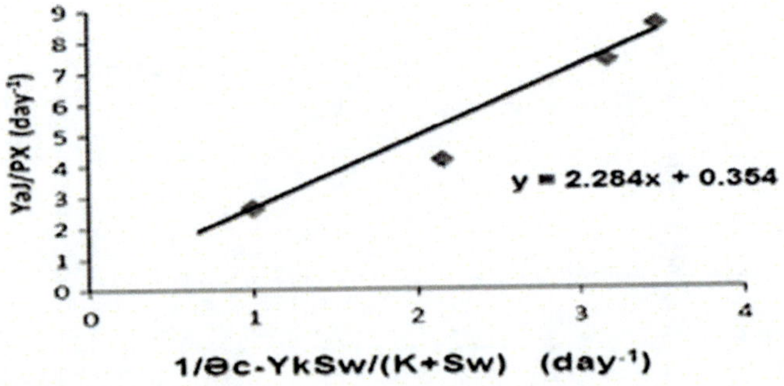

Fig. 8.6 Biomass balance for suspended growth for the determination of b_s, b_t and b_d (for batch study with municipal wastewater)

Accordingly, a best-fit line was drawn for this scatter diagram as shown in Fig. 8.4 of Annexure VI-a with a relationship $y = -0.021x + 0.129$. Considering $p = 0.93$, the value of k was determined from the intercept value. In addition, the yield coefficient, i.e., Y was obtained from the relationship $Y = \mu_{max}/k$.

Now, from Fig. 8.4 of Annexure VI-a, 1/pk = 0.129 day, p = porosity = 0.93. Hence, $k = 8.33$ day^{-1}, $Y = \mu_{max}/k = 6.36/8.33 = 0.76$.

The value of Y showed a higher value in comparison with that for municipal wastewater, as stated in earlier literatures, presumably exhibiting high rate of organic conversion.

8.2.2.3 Determination of B_s, B_t and B_d

In order to determine b_s, b_t and b_d, the biomass balance for hybrid bioreactor system at steady-state condition was considered. It was represented by the equation $YAJ/PX = 1/\theta_c - (YkS_w)/(K_s + S_w))$ (Ref. Clause 3.2, Chapter 3), where J=substrate flux within the attached-growth system, θ_c = solid retention time (SRT).

Now, θ_c is calculated as $p^* \theta$ for the case of no recirculation. The values of $(1/\theta_c - (YkS_w)/(K_s + S_w))$ and $(YaJ)/(pX)$ are presented in Table 8.2.

Accordingly, a best-fit line was drawn as shown in Fig. 8.6. The slope of the best-fit equation, i.e., $y = 2.284x + 0.354$, represents the values of $(b_s + b_d)/b_s$. On the other hand, the intersection of Y-axis represents the value $b_d(b_s + b_d)/b_s$. The coefficient of total biomass loss, i.e., b_t, can be calculated as $b_t = b_s + b_d$. From Fig. 8.6, $m = 2.284$ and $c = 0.354$, which shows biomass decay loss is 0.15 day^{-1}, biomass shear loss is 0.12 day^{-1} and total biomass loss is 0.27 day^{-1} (Ref. Sect. 3.2, Chapter 3).

Therefore, the kinetic coefficients and physical data used for the prediction of the effluent substrate (COD) concentration under continuous study with municipal wastewater are as follows:

$k = 8.33$ day^{-1}, $Y = 0.76$, $K = 0.012$ mg/cm^3, $b_t = 0.27$ day^{-1}, $b_s = 0.12$ day^{-1}, $b_d = 0.15$ day^{-1}, $D_f = 0.64$ cm^2/day (assumed).

8.3 Plotting of Predicted and Observed COD Concentration

The output data like effluent substrate concentration (i.e., COD) were experimentally obtained in continuous study with both synthetic carbonaceous and municipal wastewater. The experimental COD concentration was then compared with that derived from computer programming for establishing the accuracy of the present model. To accomplish this, the observed COD concentrations were plotted in Y-axis with respect to the COD concentrations predicted from the model in X-axis. A 45° line as well as two lines representing (+/−) 10% deviation was also drawn to understand the closeness of the experimental and predicted data. The graphical presentation of this comparison is shown in Figs. 8.7 and 8.8.

Table 8.2 Determination of b_s (biomass shear loss coefficient), b_t (total biomass loss coefficient) and b_d (biomass decay coefficient)) (for municipal wastewater)

Θ_c (day)	Y	a (cm⁻ 1)	J	p	X (mg/c)	k day^{-1})	K (mg/c)	S_w (mg/c)	$(1/\theta_c - (YkS_w)/(K + S_w))$ (day^{-1})	$(YaJ)/(px)$ (day^{-1})
0.1162	0.7	0.40	0.40	0.9	0.03	8.33	0.01	0.05	3.381255015	3.43
0.2325	0.7	0.40	0.40	0.9	0.04	8.33	0.01	0.01	0.666789555	2.78
0.3487	0.7	0.40	0.40	0.9	0.05	8.33	0.01	0.00	0.996795277	2.62
0.1162	0.7	0.40	0.40	0.9	0.02	8.33	0.01	0.05	3.47311828	8.5
0.2325	0.7	0.40	0.40	0.9	0.04	8.33	0.01	0.01	0.666789555	3.34
0.3487	0.7	0.40	0.40	0.9	0.05	8.33	0.01	0.00	0.996795277	2.67
0.1162	0.7	0.40	0.40	0.9	0.01	8.33	0.01	0.07	3.172882245	7.36

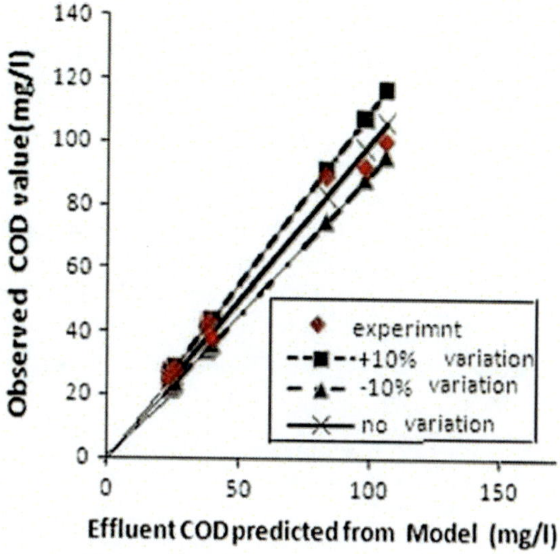

Fig. 8.7 Plotting of Predicted and Observed COD concentrations (for continuous study with synthetic wastewater)

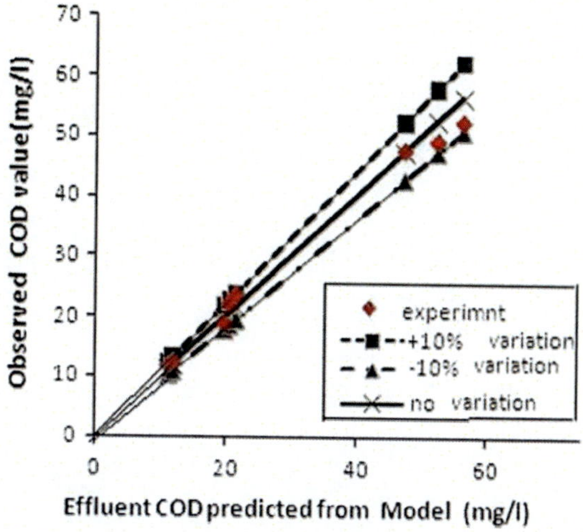

Fig. 8.8 Plotting of predicted and observed COD concentrations (for continuous study with real municipal wastewater)

8.4 Validation of Model Outputs

Followings are two cases of experiments, where the model outputs were validated:

- Case 1-continuous study with synthetic carbonaceous wastewater
- Case 2- continuous study with municipal wastewater.

8.4.1 Validation of the Model Output for Case 1 (Continuous Study with Synthetic Wastewater)

A comparison of the output results obtained from mathematical model was made with those derived from the continuous study using synthetic carbonaceous wastewater. The said comparison table is presented in Table 8.3.

It appears from the comparison table that the model output data are in a closed vicinity with the experimental observation.

Now, from Fig. 8.9, overall removal rate is calculated as 0.049 per hr, i.e., 1.176 per day. The overall removal rate observed to be quite high indicating the efficacy of hybrid bioreactor in treating carbonaceous organic matter within a short period.

Table 8.3 Comparison of exiting substrate concentration (S_W) obtained from the mathematical model with that of output from the experiment for continuous study with synthetic carbonaceous wastewater

Run No	HRT (h)	MLSS	Initial	Effluent	Predicted	%
01	4	2500	150	89	83	6.74
02		2500	200	92	98	6.5
03		2500	250	100	106	6
04	6	2500	150	42	38	9.5
05		2500	200	43	39	9.3
06		2500	250	38	40	5.26
07	8	2500	150	25	24	4
08		2500	200	27	25	8
09		2500	250	28	26	7.14

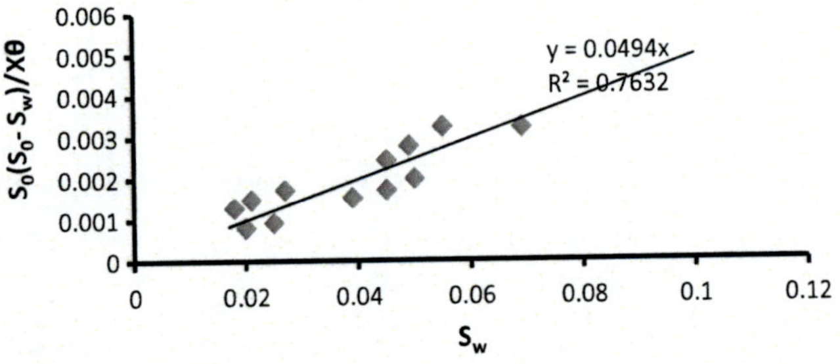

Fig. 8.9 Graph for the determination of overall removal coefficient (for continuous study with synthetic wastewater)

8.4.2 Validation of the Model Output for Case 2 (Continuous Study with Municipal Wastewater)

A comparison of the output results obtained from mathematical model was made with those derived from the continuous study using municipal wastewater. The said comparison table is presented in Table 8.4.

It appears from the comparison table that the model output data are with a little variation in respect of experimental observation. Now, from Fig. 8.10, overall removal rate is calculated as 0.054 per hr, i.e., 1.296 per day. The overall removal rate observed to be quite high indicating the efficacy of hybrid bioreactor in treating carbonaceous organic matter within a short period.

Table 8.4 Comparison of Exiting Substrate Concentration (S_w) obtained from the mathematical model with that of output from the experiment for continuous study with municipal wastewater

Run No	HRT (h)	MLSS	Initial	Effluent	Predicted	%
01	4	2000	120	47.5	47.3	0.42
02		2500	134	49	52.4	6.9
03		3000	145	52.2	56.3	7.85
04	6	2000	118	18.8	19.7	4.78
05		2500	134	22	20.5	6.8
06		3000	160	23.5	21.5	8.5
07	8	2000	110	11.7	11.2	4.27
08		2500	147	12	11.75	2
09		3000	151	12.3	11.9	3.25

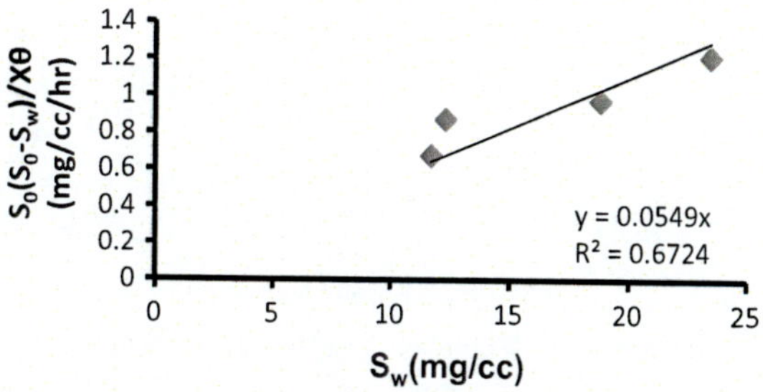

Fig. 8.10 Graph for the determination of overall removal coefficient (for continuous study with real municipal wastewater)

8.5 Process Design Approach for Fixed-Bed Hybrid Bioreactor

The process design of hybrid bioreactor is aimed to find out its volume, physical dimension and other operational requirements including oxygen for any targeted effluent substrate concentration. Besides this, it is also possible to calculate the solid retention time (θ_c), average substrate flux into the biofilm (J_{avg}) and effective as well as total biofilm thickness (L_e and L_f). There must have certain design criteria, which can be checked accordingly for any physical configuration adopted for the reactor. The process design has been performed considering steady-state condition for biomass and substrate.

Monod kinetics is assumed rational for the aerobic hybrid bioreactor treating municipal wastewater. Transport of the dissolved substrate into the liquid phase and subsequently to the attached biomass was considered due to molecular diffusion as described by Fick's 2nd law.

The rational design of the above hybrid bioreactor is contemplated employing Monod's growth pattern and associated kinetic coefficients. The values of kinetic coefficients, i.e., K, k, Y, b_t, b_s and b_d, are determined experimentally. The items for the design of an aerobic hybrid bioreactor are aeration tank capacity and dimensions, aeration facilities, porosity of the reactor, optimum hydraulic retention time (HRT), solid retention time (SRT), biofilm thickness and specific surface area of the biofilm. It is observed that economical volume of the reactor can be achieved by adopting a large value of suspended biomass (X) and a thick biofilm in the attached surface. A common range of X between 2500 and 3000 mg/L may be considered in the design of the hybrid bioreactor for the sake of good settleability. The biomass concentration in waste sludge is denoted as X_r, which may be considered between 10,000 and 12,000 mg/L, in case of hybrid bioreactor with recirculation.

8.5.1 *Determination of Volume of Aeration Basin (V)*

The volume of the aeration basin (reactor) V is calculated as per the following steps:

8.5.1.1 For Case 1: Hybrid Bioreactor Without Recirculation

Step 1: Assuming rational values of MLSS concentration, X and hydraulic retention time (θ), aJ_{avg} may be calculated for any particular value of porosity (p), as per the equation:

$$(S_0 - S_w) - \frac{pkXS_w\theta}{(K+S_w)} - aJ_{avg}\theta = 0$$

i.e.,

$$a\,J_{\text{avg(calculated)}} = \frac{S_o - S_w}{\theta} - \frac{pkXS_w}{(K+S_w)} = 0 \tag{8.1}$$

Step 2:
Now,

$$\theta_c = p^*\theta \tag{8.2}$$

Step 3: MLSS, X can now be calculated as:

$$X_{\text{calculated}} = \frac{Ya\,J_{\text{avg}}\frac{b_s}{p*b_t}}{\frac{1}{\theta c}+b_d-\frac{YkS_w}{K+S_w}} \tag{8.3}$$

Step 4: If $X_{(\text{calculated})} = X_{(\text{assumed})}$, then $V = Q^*\theta$
If

If $X_{(\text{calculated})} \neq X_{(\text{assumed})}$, then $X_{\text{calculated}}$ can be considered as the next assumed value and revised value of 'ϴ' can be obtained from Eq. (8.1) using the earlier value of '$a.J_{\text{avg}}$. Then, 'ϴ$_c$' can be calculated again from Eq. (8.2) and '$X_{\text{calculated}}$' can be calculated from Eq. (8.3). The iteration process from Step 1 to Step 4 will be continued till the requisite condition gets satisfied in Step 4.

Finally, it can be established that $X_{(\text{calculated})} = X_{\text{assumed}}$, and then, the volume of the reactor $V = Q^*\theta_{\text{revised}}$.

8.5.1.2 For Case 2: Hybrid Bioreactor with Recirculation

Step 1: Assuming rational values of MLSS concentration, X and hydraulic retention time (θ), aJ_{avg1} may be calculated for any particular value of porosity (p), as per the equation:

$$(S_0 - S_w) - \frac{pkXS_w\theta}{(K+S_w)} - a\,J_{\text{avg}}\theta = 0$$

i.e.,

$$a\,J_{\text{avg(calculated)1}} = \frac{S_o - S_w}{\theta} - \frac{pkXS_w}{(K+S_w)} = 0 \tag{8.4}$$

Step 2:

$$\theta_c = \frac{p_\theta}{\frac{Q_w}{Q}\frac{X_r}{X}} \tag{8.5}$$

Q_w waste sludge flow rate, m^3/h

X_r biomass concentration in waste sludge (mg/cc). The fraction Q_w/Q should be in the range of (0.1 – 0.5). X_r may be considered between 10,000 and 12,000 mg/L.

Step 3:
From the biomass balance for the suspended growth at steady state.

$$\frac{Ya\,J_{\text{avg}}}{pX}\frac{b_s}{b_t} - \frac{1}{\theta c} + \frac{YkS_w}{K+S_w} - b_d = 0$$

i.e.,

$$a\,J_{\text{avg(calculated2)}} = \frac{X\left(\frac{1}{\theta_c} + bd - \frac{YkS_w}{K+S_w}\right)}{(Yb_s/pb_t)} \tag{8.6}$$

Step 4: If "$aJ_{\text{avg (calculated)1}}$" = "$aJ_{\text{avg (calculated)2}}$", then $V = Q^{*}\theta$

If "$aJ_{\text{avg (calculated)1}}$" ≠ "$aJ_{\text{avg (calculated)2}}$", then aJavg (calculated)2 can be considered as the next assumed value and revised value of 'θ' can be obtained from Eq. (8.1). Then, 'θ_c' can be calculated again from Eq. (8.5) and $aJ_{\text{avg (calculated)2}}$ can be calculated from Eq. (8.6). The iteration process from Step 1 to Step 4 will be continued till the requisite condition gets satisfied in Step 4.

Finally, it can be established that "$aJ_{\text{avg (calculated)1}}$" = "$aJ_{\text{avg (calculated)2}}$", and then, the volume of the reactor $V = Q^{*}\theta_{\text{revised}}$.

8.5.2 *Determination of Surface Area of Aeration Basin (V)*

$$\text{The surface area to be provided} = (1-p)V^{*}a \tag{8.7}$$

$$p = \frac{V - V_d}{V} \tag{8.8}$$

V bioreactor empty-bed liquid volume (m^3),
V_d liquid volume displaced by the media (m^3), i.e., media volume.

Hence, V_d can be calculated for any value of p. At the same time, surface area required can be determined from Eq. 8.7.

8.5.3 *Oxygen Requirements*

Oxygen is required in the aerobic hybrid bioreactor for the endogenous respiration of the biomass in the system.

The oxygen requirement O_2 is given by,

$$O_2 = \frac{Q(S_0 - S_W)}{f} - 1.42Q_wX_r$$

f = ratio of BOD5 to ultimate BOD.1.42 = oxygen demand of biomass (g/g).

In this case, the removal of carbonaceous organic matter only has been considered.

Correction to: Fixed Bed Hybrid Bioreactor

Correction to:
S. Sarkar and D. Mazumder, *Fixed Bed Hybrid Bioreactor*,
Green Energy and Technology,
https://doi.org/10.1007/978-981-33-4546-1

The original version of the book was published without the second author. Professor Debabrata Mazumder was added as the second author retrospectively to the date of the publication and the Acknowledgements section has been updated accordingly. The correction chapter and the book have been updated.

The updated version of the book can be found at
https://doi.org/10.1007/978-981-33-4546-1

S. Sarkar and D. Mazumder, *Fixed Bed Hybrid Bioreactor*, Green Energy and Technology,
https://doi.org/10.1007/978-981-33-4546-1_9